CHINA

MAJOR WORLD NATIONS
CHINA

Rebecca Stefoff

CHELSEA HOUSE PUBLISHERS
Philadelphia

Chelsea House Publishers

Contributing Author: Tom Purdom

Copyright © 1999 by Chelsea House Publishers,
a division of Main Line Book Co.
All rights reserved.
Printed and bound in the United States of America.

3   5   7   9   8   6   4

Library of Congress Cataloging-in-Publication Data applied for

ISBN 0–7910–4735–0

CONTENTS

FACTS AT A GLANCE

Land and People

Official Name	Zhonghua Renmin Gongheguo (People's Republic of China)
Area	3,691,502 square miles (9,671,735 square kilometers)
Highest Point	Mount Everest, 29,028 feet (8,848 meters)
Major Mountain Ranges	Himalayas, Kunlun Shan, Tien Shan, Khingans
Major Rivers	Yellow (Huang), Yangtze (Chang), Huai
Major Lakes	Poyang, Dongting, Qinghai, Bosten, Songhua Reservoir
Major Deserts	Gobi, Taklimakan
Population	1,210,000,000 (1996)
Population Distribution	Countryside, 71 percent; cities and towns, 29 percent
Capital	Beijing (population 7,000,000)

Major Cities	Shanghai (population 7,830,000), Tianjin (population 5,770,000), Shenyang (population 4,540,000), Wuhan (population 3,750,000), Guangzhou (population 3,580,000), Chongqing (population 2,980,000), Harbin (population 2,830,000), Chengdu (population 2,810,000), Xi'an (population 2,760,000), Nanjing (population 2,500,000)
Official Language	Mandarin Chinese
Other Major Languages	Mongolian, Tibetan, Uygur, and dozens of Chinese dialects, including Cantonese
Ethnic Groups	Han Chinese, 92 percent; Zhuang, 1.33 percent; Manchu, .75 percent; Hui, .67 percent; Miao, .67 percent; Uygur, .58 percent; Yi, .57 percent; Tibetan, .42 percent; Mongol, .42 percent
Religions	Officially atheist. Approximately 8.3 percent of population (100 million) follow Taoism, Buddhism, Islam, Catholicism, and Protestantism.
Literacy Rate	81.5 percent of population over age 15
Average Life Expectancy	Women, 71 years; men, 68.3 years

Economy

Land Use	Meadow and pasture, 31 percent; forest, 14 percent; arable, 10 percent; other, 45 percent
Currency	Renminbi ("people's money") uses the yuan (Y) as its basic unit; 10 jiao equal 1 Y; 10 fen equal 1 jiao (about 8 Y equaled U.S. $1 in 1997).

Employment of Labor Force	Agriculture, 54 percent; manufacturing, 16 percent; trade, 6 percent; construction, 5 percent; government agencies, 2 percent; social services, 1 percent; other, 16 percent
Major Agricultural Products	Rice, wheat, corn, sorghum, millet, barley, peanuts, soybeans, sugarcane, tea, cotton, fruit, pigs, sheep, cattle, goats, water buffalo, chickens, lumber, fish, and crabs
Major Manufactured Products	Iron and steel, machinery, armaments, textiles and apparel, cement, chemical fertilizers, autos, consumer electronics, telecommunications equipment
Major Exports	Garments, textiles, footwear, toys, machinery
Major Imports	Machinery, textiles, plastics, telecommunications equipment, steel bars
Major Trading Partners	Japan, United States, Germany, South Korea, Singapore

Government

Form of Government	Republic with one legislative house and one political party
Legislature	National People's Congress
Party	Chinese Communist party
Head of Government	Premier
Official Head of State	President
Local Government	Divided into 22 provinces, 5 autonomous regions, 1 Special Administrative District (Hong Kong), and 3 cities (Beijing, Shanghai, Tianjin) that are administered by the central government and have the status of provinces

HISTORY AT A GLANCE

4000–3000 B.C.	Farming settlements exist along the Yellow River and elsewhere in north-central China. The inhabitants are the ancestors of the modern Han people.
around 2200–1700 B.C.	According to later legends, a royal family called the Xia dynasty rules the first Chinese state.
1766–1123 B.C.	The Shang dynasty governs the Yellow River valley.
1122–771 B.C.	The Western Zhou dynasty conquers the Shang territories and rules them from its capital at Xi'an.
770 B.C.	Under attack by rival states, the Zhou rulers move their capital to Luoyang. They govern for five centuries as the Eastern Zhou dynasty. Under Zhou sponsorship, K'ung Fu-tzu (Confucius) and Lao-tzu publish their writings, which form the basis of Chinese religion and philosophy for centuries. Isolated tribes populate the south and nomads from Manchuria, Siberia, and Mongolia invade the north.
5th to 3rd centuries B.C.	With the Eastern Zhou dynasty grown weak, China enters the period known as the Age of Warring States, in which many small states and kingdoms compete for supremacy.

221–210 B.C.	Shi Huangdi of the Qin dynasty conquers the Warring States and becomes the first emperor of a united China, ruling over about 40 million people. Shi Huangdi begins building the Great Wall to protect China from northern raiders. He is buried in an elaborate underground tomb complex guarded by thousands of terracotta statues of warriors.
206 B.C.–A.D. 220	Seizing control of China after the fall of the Qin dynasty, the Han dynasty adds parts of Korea, Southeast Asia, and central Asia to China. Buddhism is introduced and spreads rapidly. The historian Ssu-ma Ch'ien begins a series of annals, or historical records, that are compiled for centuries.
3rd to 6th centuries A.D.	The empire breaks up, and three kingdoms vie for power. Raids and invasions by nomads from the north and west occur often.
6th to 10th centuries	The Sui dynasty (589–618) and the Tang dynasty (618–907) reunite the empire. Crafts, trade, and the arts flourish. Digging begins on a Grand Canal to connect Beijing with the Yellow and Yangtze rivers.
960–1279	Following decades of dynastic rivalry, the Song dynasty comes to power. Gunpowder, movable type, and the magnetic compass are invented.
12th to 13th centuries	Invasions by Mongols under Genghis Khan and Kublai Khan force the Song to retreat to southern China. Kublai conquers all of China and starts the Yuan dynasty of Mongol rulers in 1279. The Yuan rulers complete the Grand Canal and receive traders and missionaries from Europe, including Marco Polo.
1368	Native Chinese overthrow the Mongol rulers and establish a new dynasty called the Ming, which rules until 1644. The Ming rulers try to limit China's contact with Europeans.

1644 Manchus from the northeast overthrow the Ming and establish the Qing dynasty, China's last imperial house. Under the Qing, Chinese territory includes Tibet, Taiwan, Manchuria, Mongolia, and much of central Asia. The Qing continue the Ming policy of isolation from foreign influences and trade.

1839–1842 China loses the First Opium War with Great Britain.

1842 The Treaty of Nanjing forces China to open some ports to European and American merchants.

1850–1864 The Taiping Rebellion, a widespread peasant revolt against the Manchus, rages throughout southern China. Before it is crushed by imperial forces, the rebellion kills 20 million people.

1856–1860 The Second Opium War, with Great Britain and France, results in another defeat for China. Foreign powers take almost complete control of the empire.

1899–1900 Nationalist societies hoping to drive foreigners out of China stage an uprising called the Boxer Rebellion. European, American, Russian, and Japanese forces unite to end the rebellion.

1912 A revolution against the Manchu imperial dynasty drives the last emperor, a child named Pu Yi, off the throne. The Chinese republic is founded, with Sun Yat-sen as its leader.

1916 The republic breaks up. The north is governed by a group of warlords; the south is held by Sun Yat-sen's Nationalist party, called the Guomindang.

1925 The Guomindang, in alliance with the Chinese Communist party, unifies the country under the rule of Chiang Kai-shek.

1927   The alliance between the Nationalists and the Communists ends, and the two groups go to war.

1931   Japan invades and occupies Manchuria (northeast China).

1934–1935   In a trek called the Long March, Communists travel north from their strongholds in the southern province of Jiangxi. Led by Mao Zedong, they establish new headquarters in the northwest.

1937   Japan attacks China. Both Nationalists and Communists battle the Japanese, who withdraw at the end of World War II in 1945.

1949   After several years of civil war, Mao Zedong proclaims China a Communist state, the People's Republic of China (PRC), with its capital at Beijing. Mao runs the country as head of the Communist party. The Nationalists take refuge on the island of Taiwan.

1950   China enters the Korean War on the side of Communist North Korea and also invades Tibet.

1960s   China occupies some territory inside India's borders but later withdraws. Relations with the Soviet Union deteriorate, and China competes with the Soviet Union for prominence in the world Communist community.

1966   Under Mao's guidance, young Communist party members called Red Guards launch the Cultural Revolution, an attack on tradition, intellectualism, and Western influences.

1971   The People's Republic is given China's seat at the United Nations, replacing the Nationalist government of Taiwan.

1972   Richard Nixon is the first U.S. president to visit China. His visit paves the way for diplomatic and trade relations between the United States and the People's Republic.

1976 Zhou Enlai and Mao, the top party leaders for decades, die. Deng Xiaoping emerges as the new leader of China.

1978 Leadership begins move to less centralized, more market-oriented economy.

1982 China adopts its fourth constitution since the founding of the People's Republic. Deng and officials selected by him remain in power.

1989 In June, thousands of students and workers demonstrating for democratic reforms are killed by the army after refusing to disperse. The center of the protest and massacre is Tiananmen Square in Beijing.

1990 President Yang Shangkun wins control of the military command from Deng Xiaoping.

1992 Deng visits southern China and reaffirms his support for market-oriented, economic reform. Younger, reform-minded leaders receive top positions at the 14th Party Congress.

1997 In February, Deng Xiaoping dies. In July, Hong Kong again becomes part of China.

A mother carries her daughter on her back so that she can use her hands for work in the Emei mountain region of Sichuan province. In 1982, China became the first country to have more than 1 billion inhabitants.

China and the World

One of every five people on earth is Chinese. In terms of China's place in the world, its immense population is perhaps the country's most significant feature. In 1982, China became the first country to have 1 billion inhabitants, and it continues to have the highest population of any country—about 1,210,000,000 people in 1996, or slightly more than one-fifth of the total world population. This fact makes China a powerful force in global politics and international trade.

China is large in size as well as in population. It covers most of the eastern part of the Asian continent and is the world's third largest nation, after Russia and Canada. Its sprawling size encompasses many types of terrain: subtropical jungles, deserts, glaciers, and desolate mountain ranges. The typical landscape in China's most crowded regions, however, is one of intensively cultivated farmland—green fields crisscrossed by countless tiny irrigation ditches, or rock-walled, terraced garden plots stacked one above another on a hillside. China's economy has been based on agriculture for thousands of years, and its farmers have learned how to harvest crops from every possible inch of fertile land.

Modern China is in the process of becoming an industrial nation, using its large reserves of coal and minerals to fuel and supply its growing number of factories, but more than half of its laborers still work in agriculture.

China is not only vast; it is also old. Archaeologists have learned that people have lived in China for hundreds of thousands of years. By 5,000 or 6,000 years ago, settled farming communities existed along the Yellow (Huang) River, which winds through a wide, fertile plain in north-central China. This was the birthplace of Chinese civilization. By the 18th century B.C.—nearly 4,000 years ago—a royal family called the Shang dynasty had achieved control over much of the Yellow River valley. (A Xia dynasty may have ruled the region even earlier, around the 23rd century B.C., but scholars have not yet determined whether or not the Xia existed in fact or merely in legend.) The Shang dynasty was followed by the Zhou dynasty, which came to power in 770 B.C. The Zhou sponsored a flowering of literature and the arts; it was under Zhou rulers that two of China's most celebrated thinkers, K'ung Fu-tzu (Confucius) and Lao-tzu, published their philosophical writings, which have long been regarded as the cornerstones of classical Chinese culture. The power of the Zhou rulers waned within a few centuries, and throughout China other kingdoms gained strength and fought for power. But in 221 B.C. a prince of the Qin dynasty conquered these warring states and became the first emperor of a united China. The Qin dynasty was succeeded by others: the Han, the Sui, the Tang, and the Song. Periods of firm central rule under powerful dynasties or exceptionally capable emperors alternated with eras of internal conflict or disruption, when no strong leader emerged to hold the empire together. Several dynasties—for example, the Yuan and the Qing—were founded by invaders from outside China who came to conquer and stayed to rule. Nevertheless, Chinese language, art, and culture remained remarkably consistent over thousands of years.

To farm the land in an area of Shanxi province that is periodically devastated by floods and droughts, the Chinese must terrace the fields. The country has to cultivate as many fields as possible to grow food for its enormous population.

That cultural consistency was caused, at least in part, by China's isolation from other centers of civilization, especially from the Middle Eastern and European cultures that grew up around the Mediterranean Sea. Although the people of ancient and medieval China did have some contact with Europe—including the visit by Marco Polo, an Italian merchant, to the imperial Chinese court in the 13th century—such contact was limited by the great distance that separated China from the West and by the difficulties of travel. Then, in the second half of the 14th century, when Europe stood on the threshold of an era of exploration that would result in the "discovery" of the Americas and the dramatic acceleration of world travel and trade, the Ming dynasty came to power in China. The Ming wanted to keep Chinese culture and politics free of foreign

20

In 1982, a musician from Inner Mongolia plays a traditional instrument called a hu-ch'in, which is similar to a four-stringed fiddle. Chinese language, music, art, and culture have remained extraordinarily consistent over thousands of years.

influences, so they established a policy of isolation from the rest of the world that kept China's borders largely closed to foreigners for 500 years.

This period of self-imposed isolation ended in the mid-19th century. At that time China's imperial dynasty was weakened by internal conflict, and foreign powers such as Great Britain and the United States took advantage of this weakness, using their superior military strength to gain favorable trade agreements and a considerable measure of political influence over the decaying empire. Around the same time, Chinese emigrants left home in large numbers, many of them to work on such projects as the transcontinental railroad in the United States. These emigrants carried their language, culture, and customs to almost every part of the world. Today nearly all major cities have Chinese communities.

In modern times, China's history has been shaped by two pivotal events. The first was a revolution that overthrew the last imperial dynasty in 1912. China was proclaimed a republic, but its first few decades were consumed by civil war, as various groups within the country strove for power. The principal opponents were the Nationalist party and the Chinese Communist party. The Communists gradually gained strength by recruiting millions of members from China's vast, impoverished peasant class. By 1949, the Communists drove the Nationalists out of China and established a Communist state called the People's Republic of China (PRC). The Chinese Communist party has remained in power ever since.

For many years after the Communists founded the PRC, the non-Communist Western nations—especially the United States—refused to recognize the PRC as a legitimate government. China's seat in the United Nations was held by the Nationalists, who had fled to the island of Taiwan and set up their own government there. Eventually, however, Western leaders had to face reality: A nation of 1 billion people, whose leaders possessed nuclear weapons and technology, could no longer be ignored. In 1971, the PRC was voted into China's United Nations seat, replacing the Nationalist government of Taiwan. And in 1972, U.S. president Richard Nixon visited China, paving the way for increased communication between the two nations after a quarter of a century of silence.

During the 1970s and 1980s, relations between China and the West steadily improved. Tourists from the United States and other Western nations visited China by the thousands to see such sights as the 2,000-mile-long (3,200-kilometer-long) Great Wall, built by ancient emperors to protect China's heartland from northern barbarians. Trade agreements multiplied, and products made in China appeared in stores around the world. At the same time, opposition to the rigid rule of the Chinese Communist party was growing within China. In the late 1980s, that opposition became public as scores of dissidents (antigovernment demonstrators), many of them students, launched a series of protests and demanded that China's government be made more democratic. The protests reached a bloody climax in June 1989 when Deng Xiaoping, head of the Chinese government since the late 1970s, ordered the army to break up a large demonstration in Tiananmen Square in Beijing, the capital of the PRC. Television viewers around the world tuned in their news programs and watched in shock and distress as tanks ground their way into the square. Hundreds of demonstrators were killed and thousands more were arrested; as many as 30,000 dissidents may have been jailed. Despite widespread opposition, the Communist party remains in control as China enters a new century.

China has undergone more changes during 50 years of Communist rule than during many centuries under imperial rule. Some of the changes are undoubted improvements. Many diseases, such as smallpox, cholera, and typhus, have been brought under control, and education, once the privilege of the aristocracy, is now available to all. Women, traditionally regarded as a form of property, are now guaranteed equal rights by the constitution. Social aid programs have eliminated extreme poverty. Yet modern China faces many problems. Among them are unrest and uprisings among non-Chinese ethnic groups in various parts of the country; economic shortcomings; high unemployment in some regions; transportation and communications systems that cannot keep pace with the needs of the people; and environmental deterioration and pollution. But China's most pressing problem may be the conflict between governmental control and the demand for basic human liberties.

The Great Wall, which is 2,000 miles (3,200 kilometers) long, was built by ancient emperors more than 2,200 years ago to protect China's heartland from the raiding barbarians of the north. The wall traverses mountains and valleys and runs from the Shanhaiguan Pass along the Bo Hai to the Jiayuguan Pass in Gansu province.

2

The Land

China is the largest country in Asia and the third largest in the world. The country stretches for about 3,100 miles (5,000 kilometers) from east to west and about 3,400 miles (5,440 kilometers) from north to south. It is about twice the size of Europe, with a total area of 3,691,502 square miles (9,671,735 square kilometers). This area includes several islands; the largest of these is Hainan, off the southern coast.

China occupies most of eastern Asia. Its land frontier is 12,400 miles (19,840 kilometers) long. In the northwest, China borders nations of the former Soviet Union, especially Kazakhstan. In the northeast, China borders North Korea. Mongolia—the homeland of the Mongol peoples—is sandwiched between Russia and China. The large region that was once called Outer Mongolia is now an independent nation, the Mongolian People's Republic. A somewhat smaller region called Inner Mongolia (Nei Monggol) is part of China; it lies inside the Chinese border, just southeast of the Mongolian People's Republic. On the other side of the country, in the extreme west, China shares a very short border with Afghanistan.

Bathers spend a day at the beach near the city of Dalian on the Liaodong Peninsula. China's eastern border, which is more than 3,588 miles (5,774 kilometers) long, is seacoast.

In the southwest, it shares longer borders with Pakistan, India, Nepal, Bhutan, Myanmar (formerly Burma), Laos, and Vietnam.

The entire eastern border of China is seacoast, forming a coastline of 3,588 miles (5,774 kilometers) along the Pacific Ocean. The portions of the Pacific that wash China's shores are called the Gulf of Tonkin (Tongking) and the South China Sea in the south (around Leizhou Peninsula and Hainan Island), the Taiwan (Formosa) Strait in the southeast (between China and the island of Taiwan), the East China Sea in the east (off the coastal city of Shanghai), and the Yellow Sea in the northeast (between China and the Korean peninsula). A large bay in the Yellow Sea is called the Bo Hai (Gulf of Chihli); the city of Tianjin is located on this gulf, and China's capital city, Beijing, is a short distance inland from it. The gulf is framed by two peninsulas: Liaodong on the north and Shandong on the south. The southern part of China's seacoast, south of the city of Hangzhou, is rocky, with many bays, coves, capes, and tiny islands. Much of the northern coast, however, is flat and sandy.

Regions and Provinces

China can be divided into several very large general regions. The heartland of the country, usually called inner China or China proper, extends from the Pacific coast west to the city of Chengdu and north to Bo Hai. It was here that Chinese culture and history began thousands of years ago, and it is here that the majority of the nation's ethnic Chinese live today.

The four other general regions of China are areas that were not originally inhabited by ethnic Chinese but have been under Chinese control for many years. These four regions are Manchuria or Northeast China, Inner Mongolia, Chinese Turkestan or Chinese Central Asia, and Tibet (Xizang). Inner Mongolia, Chinese Central Asia, and Tibet are the most sparsely settled parts of China.

Because China is such a big country, its rulers have long found it convenient to divide the country into provinces that could be administered by local governors. Today, China is divided into 22 provinces, 5 autonomous regions, and one Special Administrative District. The names and boundaries of many of these provinces have been in use for more than 1,000 years. Hebei, Shandong, Jiangsu, Zhejiang, Fujian, Guangdong, and Hainan Island are located on the coast, from north to south. The westernmost provinces are Gansu, Quinghai, Sichuan, and Yunnan. The interior provinces are Shanxi, Shaanxi, Henan, Hubei, Anhui, Guizhou, Hunan, and Jiangxi. The Manchurian provinces are Heilongjiang, Jilin, and Liaoning.

The Special Administrative District is the city of Hong Kong, which Britain ruled from 1898 to 1997. The Chinese government agreed to a policy of "one country, two systems" when it regained sovereignty over the colony.

The five autonomous, or self-governing, regions were established in recent years by the Chinese government in areas where the native population was not Chinese and where traditional provinces did not exist. They are the Xinjiang Uygur autonomous region in the northwest, populated by a people called the Uygurs; the Tibet autonomous region; the Inner Mongolia autonomous region; Ningxia, or the Huizu autonomous region, south of Inner Mongolia, populated by a people called the Hui; and the Guangxi Zhuangzu autonomous region in the south, between Yunnan and Guangdong provinces, populated mostly by a people called the Zhuang. In addition to these five regions, many small autonomous minor regions and counties are scattered through the provinces wherever the population contains a high percentage of people who are not ethnic Chinese.

Three important districts are not included in either the provinces or the autonomous regions. They are the large cities of Beijing, Shanghai, and Tianjin. These cities are administered by the central government and have a status similar to that of provinces.

From Lowlands to Mountain Peaks

China's physical geography is immensely varied, ranging from the icy peaks of Tibet's Himalaya Mountains to the arid deserts of the northwest; from the cool, windswept plains and forests of Manchuria and the flat, fertile river valleys of the middle provinces to the subtropical jungles and terraced rice fields of the south. China can claim the highest point on earth: Mt. Everest (called *Qomolangma* in Chinese and meaning "goddess mother of the world"), which reaches an altitude of 29,028 feet (8,848 meters) and straddles the border between Tibet and Nepal. China also contains one of earth's lowest points: Lake Ai-ting—a saltwater lake in a lowland called the Turpan Depression in the northern part of Xinjiang Uygur—is 505 feet (153 meters) below sea level.

China rises in three levels of elevation from east to west. The first level is the eastern lowlands—broad, flat valleys and river deltas that are the agricultural heart of the country. These lowlands make up about 20 percent of China's total territory. Although the coast is hilly and rugged in places, especially in the south, most of the coastal region is below 1,500 feet (455 meters) in altitude.

The second level covers about half of China. It rises from the coastal lowlands into a jumbled mixture of mountain ranges, steep and narrow valleys, plateaus, and wide, level basins surrounded by mountains. This is the terrain of the central, western, and southern provinces, of Inner Mongolia, and of the Xinjiang Uygur autonomous region. Altitudes in these parts of China range from 3,000 to 6,000 feet (909 to 1,818 meters).

In the southwest is China's third and highest level: the Plateau of Tibet (Xizang Gaoyuan), which occupies all of the Tibet autonomous region and most of Qinghai province—more than 25 percent of China's total area. The northern part of the plateau, in Qinghai province, is called the Qaidam Basin; it consists mostly of desertlike areas of gravel, sand, and salt, with a large lake of salty water in eastern Qinghai. South of the Qaidam Basin is the central part of the Plateau of Tibet, a cold, high desert with an average altitude of more than 13,000 feet (4,000 meters). Lhasa, the capital

The grasslands of Inner Mongolia provide the perfect environment for raising horses. Inner Mongolia, which is located in northern China, has a terrain that includes plateaus, dry steppes, vast areas of desert, large basins of inland drainage, and tall mountain ranges.

of Tibet, is almost 12,000 feet (3,636 meters) above sea level. The far western part of the plateau, called the Changtang, is more than 16,500 feet (5,000 meters) above sea level and is sometimes called "the roof of the world." The Himalayas form the southern edge of the plateau and China's southwestern border.

About a third of China is covered with mountains. The major mountain ranges other than the Himalayas are the Tien Shan, or "Celestial Mountains," which extend from northwest China into the former Soviet Union; the Kunlun Shan (*shan* means mountains), which separate the Plateau of Tibet from the Xinjiang Uygur autonomous region; the Greater Khingan (Da Hinggan) and Lesser Khingan (Xiao Hinggan) ranges of Manchuria; the Altai Shan, which form the border between Xinjiang Uygur and the Mongolian People's Republic; the Qin Ling (*ling* means mountain range in Chinese) of central China proper; and the many lesser but still formidable ranges that separate China proper into a multitude of isolated basins and valleys, especially in the south and west.

This tumbled, mountainous terrain has shaped China's history. Chinese culture and civilization developed in the fertile plains and valleys along the Yellow River, and this region was the political center of China for centuries. Other parts of the country were later absorbed into the Chinese empire, but the central government sometimes found it hard to gain complete control over the outlying regions, which were separated from the center by mountain ranges, gorges, and rivers. Many provinces and districts—such as Sichuan, a fertile basin in the southwest, about twice the size of Scotland and cut off from the rest of China by a ring of forbiddingly steep mountain ranges—maintained a high degree of independence, preserving their own customs, cultures, and even languages. Throughout China's history, only exceptionally strong or efficient governments were able to keep these isolated frontier provinces from splitting off into independent kingdoms.

Navigable waterways such as this river in Jiangsu province are important for transporting goods and people. China has more than 50,000 rivers, most of which are located in its eastern zone.

Physical Characteristics

Geographers divide China into eastern, northwestern, and southwestern zones. The chief feature of the eastern zone is its many rivers. China has more than 50,000 rivers, and in the east they have carved countless ravines and canyons. They have also deposited deep layers of soil to form fertile river valleys and deltas. The southwestern zone is high and cold; its chief feature is mountains. The northwestern zone is extremely dry; its chief feature is the wind, which erodes soil, carries away moisture, and creates deserts of shifting sand dunes. Each of these zones contains many distinctive areas.

The eastern zone includes most of the provinces. The northernmost part of this zone is Manchuria, which is ringed by the Khingan ranges. Central Manchuria is a huge plain covered with forests, meadows, grassy steppelands, and farms. Watered by the Songhua and Liao rivers, the Manchurian plain is an extremely productive agricultural region. A reservoir built on the Songhua River has become one of China's largest lakes. South of the Manchurian plain are a series of low but rugged mountains called the Changbai Shan, or "Forever White" mountains, because of the snow on their peaks; their lower slopes are heavily forested. This region produces timber, animals that are valuable for their fur, and herbs that are used in traditional Chinese medicine.

Much of the eastern zone is occupied by the North China Plain, the heartland of Chinese civilization, a low and almost perfectly flat region along the Yellow and Huai rivers. This plain is the most densely cultivated and populated part of China. Just northwest of the plain, the Shanxi, Shaanxi, and Gansu provinces form a vast plateau made of loess, a fine yellowish gray soil that is very fertile. Loess has nourished Chinese agriculture for thousands of years. It is carried down to the North China Plain by the Yellow River (which gets its color from the loess) and hundreds of smaller rivers. After millennia of erosion, at least half the surface of the loess plateau is furrowed by gulleys and ravines, some of them 650 feet (200 meters) deep. South of the loess plateau and west of the North China Plain is the Qin Ling, the largest range east of Tibet. Its highest point, Taibai Shan, is 13,474 feet (4,083 meters) above sea level.

Another important mountain in the eastern zone is Tai Shan, on the Shandong Peninsula. Although it is not a particularly high peak—its altitude is 5,048 feet (1,530 meters)—Tai Shan has been regarded as a sacred mountain since the dawn of Chinese civilization. It is said that the first kings of China made pilgrimages to Tai Shan and performed holy sacrifices there to ensure that their rule would continue. The philosopher K'ung Fu-tzu lived and wrote on the slopes of Tai Shan in the 6th and 5th centuries B.C. Over thousands of years, many temples and shrines were built along the road leading to the peak; some of them still stand.

South of the North China Plain is another fertile, densely populated region, the valley of the Yangtze (Chang) River. The longest river in China, the Yangtze runs for nearly 4,000 miles (6,400 kilometers) from its source high in the western mountains to the wide mouth where it flows into the East China Sea. Far inland from the sea, east of the city of Yibin (in Sichuan), the Yangtze passes through a stretch called the Three Gorges, where the river is hemmed in by towering cliffs and races over a series of dangerous rapids. Closer to the sea, the river runs past two of China's largest lakes, Poyang and Dongting. The cities of Nanjing, Shanghai, and Hangzhou are located in the Yangtze delta.

South of the Yangtze, still in the eastern zone of the country, is an area of mountains cut by short, swift rivers. In the steep valleys and on narrow strips of flat coastland, farmers of this region have mastered the art of using rock-walled terraces to prevent soil erosion and to use every inch of level ground. The rocky ocean shoreline is dotted by hundreds of small islands, most of them inhabited by fishermen and their families. Inland from this south-

The limestone mountains, or karst, of Guangxi province, near Yangshuo, are characterized by underground streams and caverns. The fertile rice fields of the area are watered by the Jingbao River, which snakes around the majestic peaks.

A camel caravan winds its way down the cliffs of the Gobi Desert in northwestern Inner Mongolia. The Gobi Desert is from 3,000 to 5,000 feet (910 to 1,520 meters) high, has cold winters and short, hot summers, and often has brutal wind-and-sand storms.

eastern coastal region are the Nan Mountains. The Pearl (Zhu) River flows out of these mountains into a plain on the south coast. The city of Guangzhou (formerly called Canton) is located at the mouth of the Pearl River, as is the former British colony of Hong Kong, which became part of China in 1997. North and west of the Nan Mountains are Hunan and Sichuan provinces. The tallest peak in China proper is in Sichuan. It is called Gongga Shan, and it is 24,935 feet (7,556 meters) high.

China's southwestern zone includes the Guangxi Zhuangzu autonomous region, the provinces of Guizhou and Yunnan, and the Tibet autonomous region. The Guangxi region and the two provinces consist mostly of mountains that rise toward Tibet. These ranges are cut by canyons and occasionally broken by wide upland plateaus; some small lowlands exist along the rivers. One of China's most scenic districts is in the northern part of the Guangxi region around the city of Guilin. Here the erosion of limestone mountain ranges over many thousands of years has created cone-shaped, isolated peaks that rise abruptly from the level plain that surrounds them. This region of limestone landforms is called karst and is characterized by underground streams and caverns. Shaggy with vegetation, swathed in mist, these unusual mountains have been portrayed in Chinese paintings and drawings for centuries. Some of them are topped by temples or inns that can be reached by steep, winding paths. They have traditional names, such as "Folded Brocade Mountain," said to have been named by an empress who ordered her weavers to produce a roll of satin cloth to match the color and texture of the tree-covered peak. The provinces of Guizhou and Yunnan are scenic as well, with many impressive waterfalls and dense jungle like that of Southeast Asia.

The northwestern zone of China is Chinese Central Asia. It includes Inner Mongolia, Qinghai province, and the Xinjiang Uygur autonomous region. The major cities are Hohhot, Xining, and Ürümqi. This arid zone contains the Gobi Desert (which stretches across the border into the Mongolian People's Republic), the Qaidam Basin, and the Turpan Depression. The westernmost part of the zone is the Tarim Basin, which lies between the Tien and Kunlun mountain ranges. This flat wasteland of mixed sand and gravel covers about 215,000 square miles (563,300 square kilometers). At the center of the Tarim Basin is the Taklimakan Desert, one of the driest places on earth, with almost no plant or animal life. North of the Taklimakan, the Tien Shan are well watered by many streams. The slopes and valleys of these mountains offer excellent grazing and have been roamed by nomadic herders for thousands of years. Glacial lakes are found among the peaks; Bosten Lake is the largest. North of the Tien Shan and the Turpan Depression, in the northwestern corner of China, is a pasture-covered lowland called the Junggar (Dzungarian) Basin. Leading westward out of Mongolia like a grassy highway, Junggar was for hundreds of years the route along which eastern peoples such as the Huns and the Mongols launched their raids on Russia and Europe.

Climate and Weather

Most of China has a temperate climate—that is, cold winters and hot summers, with well-marked changes of season. The mountainous regions are colder than surrounding lowlands. The southern coast is subtropical and has no true winter.

Except for the highest parts of the Himalayan range, northern Manchuria is China's coldest region, with a mean annual temperature below freezing; winter temperatures can go as low as -17 degrees Fahrenheit (-27 degrees Centigrade) and summer temperatures rarely go above 54 degrees Fahrenheit (12 degrees Centigrade). Winter brings snowfall to the northern half of China, about as far south as the Qin Ling. Most of China proper is very warm in the summer, with the hottest temperatures to be found in the Yangtze River valley, where the mean July temperature is between 84 and 86 degrees Fahrenheit (29 and 30 degrees Centigrade). The warmest region year-round is the far south, in Guangdong province. The mean annual temperature of the Pearl River valley is about 68 degrees Fahrenheit (20 degrees Centigrade).

Rainfall occurs mostly in the summer and is greatest near the ocean coast. The southeastern part of the country receives the highest rainfall, with an average of 80 inches (200 centimeters) a year in the city of Guangzhou. The northern and northeastern regions—around Beijing, and in Manchuria—receive about 25 inches (60 centimers) of precipitation a year, whereas the Plateau of Tibet and the northwest receive less than 4 inches (10 centimeters). Extremely arid regions, such as the Taklimakan Desert, may receive no rain at all in some years.

Along with the summer rains, which can drop as much as 12 inches (30 centimeters) of rain in a single day on the southern coast,

The Temple of the Fragrance of Buddha sits atop the Hill of Longevity (on Lake Kun-ming) at the Summer Palace (Yiheyuan) in Beijing. The Summer Palace is an immense tour de force in landscape gardening and receives approximately 25 inches (60 cen-timeters) of precipitation annually.

come summer storms. Typhoons, or tropical cyclones, are rain-laden high winds that lash the coast, most often in August and September.

Plant and Animal Life

Much of China—especially the lowland river valleys within China proper—has been the home of a large farming population for centuries. People have used the land and thereby changed it, re-placing wild vegetation with crops and native wild animals with domestic farm animals. Forests in particular have suffered; at one time, at least half of China was covered with trees, but centuries of clearing land for farms and burning wood for fuel have reduced the forested area to about 14 percent. Despite widespread intensive cultivation, some isolated parts of China proper remain in their natural state. And large expanses of Chinese Central Asia and the Tibetan plateau are almost untouched by settlement or develop-ment.

38

China has one of the world's most diverse collections of native plant and animal life. Nearly all of the major types of plants found in the Northern Hemisphere can be found in China, which has more than 30,000 species of seed plants and more than 7,000 species of woody plants, including trees. Some trees are economically important. Tung trees, camphor trees, lacquer trees, and star anise trees yield oils that are used in furniture polishes, medicines, and foods. Larch, pine, birch, and other timber trees are harvested in the northeast. The rain forests of Hainan Island and southern Yunnan province contain valuable hardwood trees, such as mahogany, that are prized by furniture makers. Several trees are found only in China. The metasequoia and the cathaya are 60 million-year-old species that long ago died out elsewhere in the world. China also has thousands of varieties of flowering plants, including 400 types of peonies and 650 types of azaleas.

Vegetation in China follows the pattern of climate, ranging from the mangrove swamps and palm groves of the far south, through the evergreen forests of the central provinces, to the hardy desert shrubs and grasses of the Tarim Basin in the northwest. In general, the southeastern part of China is known for its forestland, whereas the northwest has long consisted of mixed grassland and desert. The northeast has both oak forests and grasslands.

One plant in particular is often associated with China. Bamboo is a woody type of grass that grows in the form of a hollow, jointed cane and reaches heights of 20 feet (6 meters) or more. It flourishes south of the Yangtze River, and for centuries its stems have been used to make furniture, flutes, and household objects. Bamboo is vital to the survival of China's best-known native animal, the giant panda, which feeds only on bamboo shoots. These large black-and-white mammals that are related to raccoons but resemble bears are now found only in remote parts of Sichuan province and eastern Tibet; scientists estimate that about 1,000 of them survive in the wild.

Native Chinese wildlife also includes a number of rare or endangered species: the takin (a mountain-dwelling antelopelike goat), the lancelet (also called amphioxus, an ancient species of translucent fish that lives only in Fujian province), the northeast China tiger (found in Manchuria), the snow leopard of Tibet, the Siberian white crane and the Oriental white stork, the Asiatic elephant, the white bear found only in Hubei province, the mandarin duck (a small, brilliantly colored bird of the northeast), the golden-haired monkey, and the Chinese river dolphin. More common animals include deer, hares, monkeys (in the south), and more than 1,000 species of birds. China's rivers are rich in fish life, especially carp and catfish.

China faces serious environmental problems. Some of these problems, such as air and water pollution, are caused by the growth of industry. Others, such as soil erosion and the destruction of forests for fuel, are caused by the pressure that China's huge population places on the land. Since the early 1970s, the Chinese government has enacted laws to correct—or at least control—some of the worst environmental problems. Unfortunately, these laws are not always thoroughly enforced. But at least 763 nature preserves have been created to protect China's surviving wildlife and vegetation. Several, such as the Changbai Mountain Nature reserve in the northeast, serve as bases for international research projects. The Dinghu Mountain Reserve in Guangdon province protects a subtropical evergreen forest that has remained untouched for more than 400 years, and the Nangun River Reserve in Yunnan province consists of a tropical rain forest. The best-known of China's nature parks is Wolong Reserve in Sichuan province, where many of the giant pandas live. Scientists from around the world have been invited to Wolong to study the panda; their research is aimed at finding ways to protect and preserve the species.

This life-size terra-cotta soldier (circa 210 B.C.) was among the more than 8,000 that were unearthed in 1974 when archaeologists discovered Shi Huangdi's tomb near Xi'an. Each statue is incredibly lifelike and has distinct facial features.

3

Early History

Ancient Chinese legends say that the first civilized communities in China were ruled by three kings: Fu-hsi, Shen Nung, and Huangdi. The first of these kings taught people how to hunt; the second gave them grain and invented the plow; and the third gave them fire, cleared the plains of trees, and taught them music. All three of these kings were said to have worshipped the gods atop Tai Shan. The kings were regarded as the founders of the Chinese royal houses and revered as ancestors by later kings. Legend also says that two royal houses, the Yu and the Xia dynasties, ruled China some 5,000 years ago. The earliest royal house for which solid historical evidence exists, however, is the Shang dynasty, which established kingship over the farming communities along the Yellow River in north-central China in 1766 B.C. By that time, much of China had been settled for centuries, and the Chinese language was already well developed.

The first capital of the Shang was in Zhengzhou. Modern archaeologists have excavated a walled town there, complete with workshops and tombs. Later Shang kings ruled from Anyang, which is also the site of many archaeological finds, including bronze and tortoiseshell objects with written inscriptions. Kings were buried in elaborate royal tombs in Anyang, accompanied by slaves, soldiers, horses, and even dogs, all decapitated to serve their masters in the afterlife. The Shang warriors rode in chariots and fought with spears and arrows. Skilled astronomers in Anyang developed an accurate calendar and kept records of eclipses. Craftsmen created jewelry, pottery, and household wares ranging from hairpins to fishhooks. Nobles built elaborate timber palaces. Shang traders traveled so widely in search of goods that the term *shang jen* (Shang man) was used for merchants from the Gobi Desert to the Malay Peninsula (the southernmost portion of the mainland of Southeast Asia, consisting of parts of Myanmar, Thailand, and much of Malaysia).

In 1122 B.C., the Shang were conquered by King Wu of Zhou, a state on the western frontier of Shang territory. The Zhou destroyed Anyang and set up a new capital at Xi'an. From there Zhou kings ruled much of the Yellow River valley for three and a half centuries; historians call this era the Western Zhou dynasty. The Zhou established a complex aristocratic structure that was similar to the feudal kingdoms of medieval Europe: Numerous vassal lords swore loyalty to the king, and in return they were granted considerable power in their own localities and allowed to set up courts of their own. The state also supported a large bureaucracy, with officials in charge of agriculture, roads and bridges, the judicial system, prisoners, horses, scribes, and so on. But in spite of this elaborate system of government, the Zhou rulers were not able to hold their kingdom together. Attacks by rival states, as well as power struggles within the court, weakened the Zhou. The last Western Zhou king was murdered in 771 B.C. His family fled east to the city of Luoyang, where they established a dynasty that historians call the Eastern Zhou. The Eastern Zhou ruled over a fairly small area and had little influence over neighboring states. Meanwhile, the rest of the former Zhou territories, as well as other more southerly districts, broke up into many small petty kingdoms that were constantly at war. Around the 6th century B.C., about 170 such states existed in China proper. The far south was populated by isolated tribes who were only gradually brought into contact with the mainstream of Chinese civilization, and the northern borders were under constant attack by the nomadic peoples of Mongolia, Siberia, and Manchuria.

Although the Eastern Zhou kings lacked political power, they ruled a state in which literature and the arts flourished. Scores of writers produced poetry; historical annals; volumes of laws; and books about music, mathematics, medicine, gardening, and many other subjects. Two philosophers who are regarded as China's greatest thinkers developed and published their ideas in the Eastern Zhou era. One was K'ung Fu-tzu, who lived from 551 to 479 B.C. and whose writings on ethics shaped traditional Chinese thought. K'ung Fu-tzu emphasized the importance of order and stability and the individual's responsibility to family and to the state. For K'ung Fu-tzu, the highest virtues were moderation, education, and respect for the social order. This train of thought, usually called Confucianism by Westerners, became one of the chief ingredients in Chinese religion and philosophy. Another 6th-century-B.C. philosopher, Lao-tzu, is generally regarded as the founder of the Chinese mystical philosophy and religion called Taoism; the book *Tao-te ching* (The Way and Its Power) reflects his teachings, although he probably did not write it. The goal of Taoism was personal peace and enlightenment through a simple, harmonious way of life; the relationship of the individual to society was dismissed as unimportant or even distracting. Close association with Buddhism resulted in the addition of monasteries and temples to Taoism. Ancient Chinese gods, hermits, and immortals were added to the Taoist pantheon and were often worshiped for their favors. In later years, Taoism declined into a religion of superstition, magic, and pleasure.

The later centuries of the Eastern Zhou dynasty saw important developments throughout China. Horseback riding, the use of iron, and glassblowing became common; silk making developed into a sophisticated industry, especially in the Shandong Peninsula. Politically, however, China was torn by war and rivalry among its many kingdoms. The period starting around 450 B.C. is known as the Age of Warring States. First one then another of the local dynasties sought to achieve control over the rest by snapping up their neighbors. By 403, only seven states remained. The Qin dynasty was the most powerful of these. One Qin prince was a brilliant military commander who conquered rival kingdoms, according to an ancient historian, "like a silkworm devouring a mulberry leaf." By 221 B.C. this Qin ruler had conquered the last of the warring states and could proclaim himself the first emperor of all China. In fact, it is from the name of the Qin dynasty (pronounced Chin) that the word "China" evolved.

Imperial China

Shi Huangdi was the Qin prince who became the first emperor. He disbanded the feudal aristocracy and divided his territory—an empire of about 40 million people, from north of the Yellow River to south of the Yangtze—into 36 military commanderies, or districts. He also ordered the building of a Great Wall for protection against northern "barbarian" raiders. The wall, which incorporated many smaller barricades that had been built earlier by the feudal states, took 12 years to complete. As many as 400,000 slaves died of hunger or exhaustion; it is said that some workers were killed and enclosed in the wall as sacrifices to give it strength. Shi Huangdi believed that China needed a strong central government. In 213 B.C., to prevent opposition to his rule, he had his soldiers burn all books that did not agree with the official Qin philosophy of strict imperial government and universal obedience to the law.

Shi Huangdi died in 210 and was buried near Xi'an in a huge, elaborate underground tomb complex that took 700,000 men 36 years to build. Legend says that his tomb, which has not yet been fully excavated, is a palace filled with treasure and artworks. But in 1974 archeologists uncovered one spectacular feature of the burial: an "army" of more than 8,000 life-size soldiers made of terra-cotta, a brownish orange clay. Each has a different face, and many carry bronze and wooden weapons or hold the reins of life-size war-horses also made of clay.

The Qin dynasty did not survive Shi Huangdi. For a few years after his death, his generals scrambled for power. In 206 B.C. one of them emerged victorious from the struggle and founded a new imperial house, the Han dynasty. Under Han rule, new territories became subject to China, including the region around the Gulf of

Guardians and warriors protect a carved, seated Buddha (not in photo) in the Longmen Caves near the city of Luoyang. Sculptors carved more than 100,000 images in the 1,300 caves that are located along the Yi River. The Tang dynasty (A.D. 618–906) became known for tolerating foreign religions such as Buddhism (which had originated in India) and for supporting art. The sculpture of the Longmen Caves represents the high point of Buddhist culture in China.

Tonkin, part of Tibet, northern Korea, and the Tarim Basin. One of the greatest Han emperors, Wu, who reigned from 141 to 87 B.C., succeeded in subduing the Huns, a tribe of horsemen from beyond the Great Wall who had plagued northern China for years. Around the same time, contact was made with other cultures—including India, Persia, and even the Roman Empire. One traveler in the 2nd century B.C., an imperial emissary named Chang Ch'ien, visited India and much of Central Asia and wrote an account of his travels that was widely read in Han China.

Travel, trade, and expansion brought the Chinese into contact with Buddhism, a religion that had originated in India and was spreading throughout Asia. Buddhist monks came to China, where their faith quickly became popular. Before long, Buddhism was as widespread and as much a part of Chinese thought as Confucianism and Taoism. Education and the literary arts also flourished under the Han. Schools and teachers multiplied; by the middle of the 2nd century A.D., there were more than 30,000 students in the capital city of Luoyang alone. Paper was invented, and the first dictionary was published. Two of China's most important historians, Ssu-ma Ch'ien (187–45 B.C.) and Pan Ku (A.D. 32–92), began a series of historical annals that would be continued by later scholars for many centuries. Ssu-ma Ch'ien wrote the first general history of China from its earliest times, and Pan Ku documented the Han era.

The Han dynasty lost power in A.D. 220, when the empire broke up into three kingdoms, Wei, Shu, and Wu. Over the following centuries, north China suffered many invasions from Huns and other northern and western peoples who had no difficulty getting through, around, or over the Great Wall. During this time of disruption, Chinese culture and society remained most stable in the south, where a series of short-lived states and dynasties rose and fell along the Yangtze. Most of them had their capital at Nanjing. Although the period after the Han era was one of disunion and political turmoil, it was also a time of progress in medicine, mathematics, and geography. Buddhism gained great popularity, and Buddhist monasteries and temples sprang up everywhere, especially in the Yangtze region. Trade with India and Southeast Asia increased.

In 589, the empire was reunited under the Sui dynasty. Sui rulers maintained two capitals, Luoyang and Xi'an. They also began construction on many inland waterways, the biggest of which was the Grand Canal, which was intended to link the Yellow and Yangtze rivers with Bo Hai. The Sui dynasty came to an end before the canal was completed; however, rebellion against the second Sui emperor broke out in 613, and in 618 he was murdered. One of his generals seized power and established the Tang dynasty, ushering in an era of foreign conquest and strong central rule.

The Tang empire reached its zenith under the second Tang emperor, Li Shih-min, who led Chinese armies to victory in Mongolia and across the Tarim Basin. Before his death in 649, princes from as far away as Samarkand and Bukhara—cities in what is now the neighboring state of Uzbekistan—acknowledged Chinese overlordship. At the same time, Chinese monks and scholars carried Buddhism to Japan, and foreign missionaries—including some Christians from Turkey and Syria—preached their faith in China. Confucianism flourished, and a civil service bureaucracy was created in which students had to pass grueling examinations in the Confucian classics to win government jobs.

Once the frontiers of the empire were secure, emperors and courtiers took increased interest in the sciences. One scientific project, from 721 to 725, involved the building of nine observatories in a line between Vietnam and the Great Wall to make astronomical measurements. Under Emperor Hsuan Tsung, who ruled from 712 to 756, the wealthy and leisured classes patronized the arts, leading to a burst of activity in literature, painting, and calligraphy (fine ornamental writing). Many poems and novels were written during the Tang period, and some of China's finest artists produced paintings of landscapes and court scenes on silk scrolls.

The Tang dynasty began to crumble in 755, when army generals led an unsuccessful but destructive rebellion. A few years later, Tibetans from the west and Uygurs (a Muslim people related to the Mongols) from the northwest invaded China. Their armies sacked and looted the two capitals before they were driven out. Although the Tang rulers returned to Xi'an, their hold on power had grown weak. Various parts of the empire began to break away from the central government, starting with Sichuan in 890. In the early 10th century, China was torn by conflicts. At least five royal houses squabbled over the capital cities. Elsewhere, numerous rival states were once again vying for supremacy. A tribe called the Khitan, related to the Mongols, invaded the area around Beijing and established their own state there; over time they were absorbed into the Chinese people.

Unity was restored by the Song dynasty, which rose to power in 960. A number of notable innovations appeared during the Song era; among them were paper money, movable type (printing had formerly been done from wooden blocks on which whole texts were carved), inoculations against smallpox, and gunpowder (used in cannon, bombs, and grenades). The magnetic compass was invented, and advances were made in shipbuilding and sea trade.

The Song rulers were never able to regain the firm control of the frontiers that had been achieved by the early Tang emperors. In the early 12th century, the northern part of the Song empire was invaded and conquered by the Jurchens, a people from Manchuria, and the Song capital was moved south to Hangzhou. The Jurchens were conquered in turn by the Mongol hordes led by Genghis Khan. In the south, the Song dynasty held out against the Mongol onslaught until 1279, when it was overthrown by Genghis's grandson Kublai Khan, who ruled a Mongol empire that covered most of Asia.

Kublai made his residence in China and established a new royal house, the Yuan dynasty. He built a lavish new capital city at Beijing

Kublai Khan (1215–94), the first Mongol chieftain to conquer and rule all of China, built a new capital at Beijing, completed the Grand Canal, and was a great patron of the arts and sciences.

(then called Khan-balik) and a summer palace called Shang-du north of the capital. Kublai was an energetic and efficient emperor. He completed the Grand Canal and improved the road and postal systems throughout China. Under his rule, foreigners were welcome in Chinese cities; three foreign visitors, the Polos of Venice, stayed in China for several decades, and the book that Marco Polo later wrote about his travels was most Europeans' first introduction to eastern Asia. Kublai encouraged scientific and geographical learning, and in his time China's mathematics and astronomy were the most advanced in the world. But the Yuan dynasty was greatly resented by the native Chinese, who never stopped regarding the Mongols as interlopers. After Kublai's death in 1294, no Yuan ruler of comparable strength came to the throne in China, and the dynasty grew progressively weaker. China seethed with discontent, especially south of the Yangtze, where the Mongols had allowed many Chinese bureaucrats to remain in power.

In 1368, a rebellion of ethnic Chinese toppled the Yuan and established a new dynasty, the Ming. At first, the Ming capital was at Nanjing (which means "capital of the south"), but the second Ming emperor, Cheng Tsu (who ruled from 1402 to 1424 and was also called Yung-lo), moved it to Beijing (which means "capital of the north"). This emperor also sent a series of naval expeditions—called the "treasure fleets" and commanded by an admiral named Cheng Ho—to ports throughout Asia and on the coasts of Arabia and East Africa. The purpose of these fleets was neither conquest nor trade; instead, Cheng Ho presented gifts to local rulers to impress them with the might and wealth of the Ming empire.

A 1249 woodcut depicts the different stages of mining and manufacturing salt. Kublai Khan monopolized salt production in China, and the salt tax he levied proved to be one of his most valuable sources of revenue. Marco Polo might even have served for a time as one of Kublai's tax collectors.

Unlike the Mongols, the Ming rulers after Cheng Ho had no interest in the rest of the world. They intended to keep China apart from other nations, and they gave evidence of this intention by rebuilding and extending the Great Wall. The policy of isolationism established by the Ming dynasty kept China largely out of the mainstream of world events for five centuries. But even the determined isolationism of the Ming could not keep China completely free of outside influences during an era of world exploration. Portuguese explorers landed in southern China in 1516 and managed to acquire a small colony called Macau at the mouth of the Pearl River. The Spanish arrived in 1557, and they too were able to engage in some trade on the south coast. In the early 17th century, Roman Catholic missionaries, led by a priest named Matteo Ricci, managed to acquire a foothold at the Ming court, where they were permitted to teach and preach. Around the same time, Dutch and English ships began appearing at Macau. These Europeans introduced many new products to China, some of which had come from the Americas; among them were sweet potatoes, corn, peanuts, and tobacco. Many members of the Ming court were disturbed by the growing numbers of Europeans on China's southern shores. But before the Ming dynasty could decide how to control the newcomers, it was overthrown by vigorous invaders from Manchuria.

Tiananmen (The Gate of Heavenly Peace) Square is the main gateway to the Forbidden City in Beijing. Manchu emperors forbade ordinary people to go near the gates of their winter palace, which is how the Forbidden City got its name. Today the series of buildings and courtyards are open to the public.

4

Modern History

In 1644, a rebel Chinese general-turned-bandit seized Beijing. Another general then invited the army of the Manchu, a warlike Manchurian people, into China to oust the bandit from the capital. Having accomplished this, however, the Manchu decided to stay and take over for themselves, and within a few years they controlled most of China, although a few Ming princes fled to the island of Formosa (now called Taiwan) and set up a court in exile.

The Manchu rulers founded the Qing dynasty, which was destined to be China's last imperial house. The most able Qing emperors were Kang Xi (reigned 1662–1722) and Qian Long (reigned 1736–1796). Under these rulers, China invaded Mongolia, Korea, Tibet, northern Vietnam, Burma (now Myanmar), and Nepal and received tribute and acknowledgments of overlordship from these lands. At its height, the Qing Empire probably had about 300 million inhabitants. The arts, crafts, and sciences were generously patronized by the Qing royal family; porcelain making reached the heights of refinement, and an extensive encyclopedia was published. Government censorship, however, meant that works critical of the Manchu were destroyed; some scholars were executed for antigovernment views. The Manchus allowed many Chinese officials to remain in high-level government posts, but they passed a law requiring all Chinese to shave their head, except for a long topknot. The traditional Manchu topknot thus became a symbol of oppression for the Chinese.

The Manchu controlled foreigners very closely. Foreign merchants were restricted to the ports of Macau and Guangzhou, although Catholic priests continued to be welcome at court—many of them spoke fluent Chinese and served the emperors as tutors and interpreters. The merchants, diplomats, and military adventurers of other nations, however, grew increasingly impatient with Chinese isolationism and with the elaborate system of bribes and rituals that they had to follow in order to have any contact with the Chinese court. In 1793, the British sent the earl George Macartney as envoy to the Qing court to seek more favorable trade agreements, but the mission was not very successful, and relations between China and the West grew more threatening.

After Qian Long's death in 1799, the Qing dynasty was troubled by corruption, rivalry among various branches of the royal house, and numerous rebellions. Another problem was that foreigners were selling opium in China (most of it was grown in India and imported to China by the British), despite strict Manchu laws against the use of the drug. As a result, addiction was becoming widespread among the people. In a desperate attempt to end the opium trade, the Manchu attempted to enforce new restrictions against foreign merchants and ships. The result was the First Opium War with Great Britain. It lasted from 1839 until 1842 and ended in defeat for the Chinese, who were unable to compete with Western-style gunboats and artillery. The Treaty of Nanjing (1842) forced China to open more ports to Western ships.

Around this time the Chinese peasants, particularly in the south, suffered from a series of severe famines and from raiding bandits, whom the central government seemed unable to control. These conditions led to widespread dissatisfaction with Manchu rule. In 1850, that unrest broke out into one of the bloodiest civil wars the world has ever seen, the Taiping Rebellion. The leader of the uprising was a peasant named Hung Hsui-chu'an, who claimed to have had Christian religious visions that inspired him to preach the "Tai ping Tien-kuo" (Heavenly Dynasty of Perfect Peace) and to found a new social order. He quickly mustered an ill-equipped but enormous peasant army that devastated most of China proper for a decade and a half in its attempts to overthrow the Qing.

The Society of the Righteous Harmonious Fists, called the Boxers, meet in Beijing. In 1899, the Boxers, who resented foreigners for their control of China's major businesses, railroads, and even land, led a violent revolt against foreigners and killed many until an international force put an end to the rebellion in August 1900.

While the rebellion was going on, the Manchu found themselves compelled to fight the Second Opium War, this time with Great Britain and France (1856–60); China lost again and was forced to grant the Western nations a large degree of control over its land and commerce. At this point, European troops under General Charles George "Chinese" Gordon of Great Britain stepped in to help the Manchu put down the Taiping Rebellion. Gordon was later despised by the Chinese and criticized by many foreigners because he allowed an unruly rabble of British and French soldiers to loot and burn the priceless imperial summer palace during the fighting. He did, however, defeat the Taipings at Shanghai, and the rebellion was crushed in 1864, after Hung Hsui-chu'an's suicide. It had cost 20 million lives.

By the late 19th century, China's once-mighty imperial dynasty had become weak. Russia, Japan, Great Britain, France, the United States, and Germany carved out sections of Chinese territory to administer. These nations also had a hand in managing China's railroads, finances, and trade. Many Chinese were disgusted with this state of affairs, and they began to form nationalist secret societies that were opposed to the presence of foreigners in China. One such society, called the Society of the Righteous Harmonious Fists and labelled the Boxers because its members practiced certain martial arts, organized an uprising in 1899. The Boxers killed many foreigners and besieged others in the embassy quarter of Beijing, hoping to drive all non-Chinese out of the country. Instead, in 1900 a combined force of Japanese, German, Russian, American, French, and British troops smashed the Boxer Rebellion.

In the early years of the 20th century, Russia and Japan went to war over Manchuria, and the United States helped write the treaty that settled the war; it was as though China had no say at all in what went on in its territories. Nationalist feeling was still strong in China, however, and the anti-Manchu, antiforeign groups grew stronger as the Qing grew weaker. The main leader of the nationalist

(continued on page 65)

SCENES OF
CHINA

Overleaf: *A man and a woman from the Inner Mongolia autonomous region are on their way to work. Many of the people of Inner Mongolia are animal herders and are known for their expertise in horsemanship.*

A farmer from Xi'an in Shaanxi province transports wheat in a cart. The region around Xi'an is flat, fertile, and ideal for growing wheat, cotton, and corn.

A woman cultivates a rice field near the city of Chengdu in Sichuan province. Dujiang Yan Dam, an engineering marvel that was designed in 250 B.C. and is situated northwest of Chengdu, is the irrigation system that today provides water for more than 16.5 million acres (6.67 million hectares) of land.

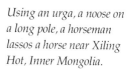

Using an urga, a noose on a long pole, a horseman lassos a horse near Xiling Hot, Inner Mongolia.

Archaeologists carefully unearth the remains of a chariot in one of the three vaults discovered at Shi Huangdi's tomb.

In 1974, some peasants who were digging a well near Xi'an accidentally stumbled upon what turned out to be thousands of life-size terra-cotta warriors standing guard by the Qin emperor Shi Huangdi's tomb. These figures were originally brightly painted and have hollow torsos, with solid arms and legs. Each face was individually sculpted.

The giant panda is China's most celebrated native animal. Although it looks like a bear, the panda is related to the raccoon and eats mostly bamboo.

The Nine-Dragon Pool is set amid Hua Qing Springs near Xi'an. An oasis of trees and red-roofed pavilions, the curative springs were discovered more than 2,800 years ago and served as a resort for China's emperors and ruling families.

The Temple of Putuozongcheng, modeled after the Potala Palace in Lhasa and built in the late 18th century, is Chengde's largest temple. Chengde, located about 159 miles (256 kilometers) northeast of Beijing and originally the Qing emperor Kang Xi's summer city, has some of the country's most unusual Chinese and Tibetan temples.

Bicyclists pass through a gate in the Huang Shan range, in Anhui province. Because of their beautifully shaped granite peaks the Huang Shan have been immortalized over the centuries by Chinese poets and artists.

The Sword Pool, surrounded by swordlike peaks that seem to pierce the sky, lies deep within the Stone Forest, which is located southeast of Kunming in Yunnan province. The Stone Forest, towering peaks of limestone, has numerous pavilions, caves, and ponds, each with its own legend.

In 1922, Sun Yat-sen, leader of the nationalist movement, posed for this photograph at his residence in Shanghai. Although he was a physician, Sun Yat-sen dedicated his life to over-throwing the Qing dynasty and establishing an independent modern republic.

(continued from page 56)

movement was Sun Yat-sen, a physician who had received some education in the West and who believed that China should become an independent modern republic.

In 1908, the Qing throne passed to an infant named Pu Yi, and Manchu nobles were appointed as regents to govern on his behalf; in reality, they were dominated by foreign interests. Pu Yi was to be China's last emperor. When the regents announced that they intended to turn control of a Chinese-owned railway over to foreign bankers, the Chinese nationalists rose up in outrage. City after city declared itself independent of Manchu rule, and soon the country was in the midst of a full-blown revolution. In early 1912, the Qing regents signed a document giving up the throne in Pu Yi's name, and the victorious nationalists declared China a republic, with Sun Yat-sen as its president.

Sun's extremely modern views were unpopular with some conservative Chinese, however, and he was soon replaced as president by Yuan Shikai, who had been a general under the Manchu. Yuan held the new republic together until his death in 1916. At that time, the Western world was convulsed by World War I and had little attention to spare for events in China. The northern part of the country, including Beijing, fell into the hands of a group of warlords. The south was held by Sun Yat-sen's Nationalist party, called the Guomindang. The Nationalists found allies in the Chinese Communist party (CCP), which had formed in response to the Communist revolution in Russia a few years earlier. Around this time Tibet and Outer Mongolia took advantage of China's preoccupation with internal troubles to declare themselves independent. Outer Mongolia associated itself with the Soviet Union and soon became an independent nation, and Tibet went its own way for several decades, untroubled by the larger powers.

Sun Yat-sen died in 1925, but his successor as head of the Nationalists, Chiang Kai-shek, managed with the help of the Communists to defeat the northern warlords and unify the country in 1925. Two years later, the fragile alliance between the Communists, who wanted to set up a new economic system based on the principles of socialism, and the Nationalists, who wanted to establish an American-style economy, broke down. The two groups went to war. At first the Nationalists had the upper hand, but the Communists patiently recruited members from the peasant class. From October 1934 to October 1935, the Communists undertook a trek they called the Long March from their strongholds in Jiangxi province north to Shaanxi province. There, under the leadership of Mao Zedong, they marshalled their forces ostensibly to fight the

Japanese, who had once again invaded Manchuria. In fact, they set up a puppet government headed by Manchu nobles, including Pu Yi.

Both the Nationalists and the Communists were opposed to the Japanese, and in 1937 they joined forces again to declare war on the common enemy. The war between China and Japan continued into World War II. Japanese forces occupied much of China, but the Chinese Nationalists received help from the Allied nations, who were fighting Japan throughout Asia. The Japanese withdrew in defeat at the end of the war in 1945, and the Nationalists and Communists promptly returned to fighting each other.

Three-year-old Pu Yi (right), China's last emperor, appears in a 1909 photograph with his father and younger brother. In 1911, the Nationalists overthrew the Qing dynasty and on January 1, 1912, Sun Yat-sen became provisional president of the republican government.

The renewed civil war between these two factions was brief but bloody. Despite support from the United States, the Nationalists steadily lost ground. Chiang Kai-shek and the Nationalists finally gave up and fled for refuge to Taiwan, as the Ming princes had done three centuries earlier. On October 1, 1949, the Communist victory on the mainland was complete. On that date Mao Zedong's Chinese Communist party announced the founding of the People's Republic of China (PRC), with its capital at Beijing.

General Chiang Kai-shek, head of the Nationalist party, reviews the Officer Training Corps in Chungking (now Chongqing) during World War II. After the defeat of Japan in 1945, the Nationalists and Communists returned to fighting each other, and in 1949, Chiang and his followers were forced to flee to the island of Taiwan.

In the following year, China entered the Korean War on the side of the Communists of North Korea and also invaded Tibet. The Dalai Lama, the head of the Tibetan Buddhist religion and the spiritual and political leader of the Tibetans, was allowed to retain control by title only. In 1959, when the Chinese army violently crushed a Tibetan revolt against Chinese rule and began the systematic destruction of Buddhist temples and schools, the Dalai Lama fled to India. Since then he has maintained a Tibetan government in exile and has sought to win world support for Tibetan independence, but China remains firmly in control of Tibet. Chinese forces also strayed across India's borders several times during the 1960s, but they withdrew after border skirmishing. China and the Soviet Union, the two giants of international communism, were friendly during the 1950s, but during the 1960s relations between them grew cold, partly because Chinese Communists refused to acknowledge the ideological leadership of the Soviets. The two nations competed for influence in the world Communist community. During the Vietnam War in the late 1960s and early 1970s, China gave support to the Communist regimes in North Vietnam and later in Cambodia.

The biggest changes in China between 1950 and 1965 were economic. Mao launched the country on a course of rapid industrialization, hoping to turn China into a self-sufficient, modern nation. His "Great Leap Forward" of 1958–59 was a set of economic reforms that was supposed to bring about a dramatic increase in the output of factories and farms. Initial success was followed by the economic crisis of 1960–62, called the "Three Bitter Years"—a period of bad harvests and shortages of raw materials caused by poor planning. The economy did not begin to recover until 1965.

The mid-1960s saw the beginning of one of the most tumultuous periods in China's history: the Cultural Revolution. From 1966 until Mao's death in 1976, zealous revolutionary brigades called Red Guards sought to impress the ideals of the Communist revolution

70

Mao Zedong proclaimed the establishment of the People's Republic of China after the defeat of the Nationalists in 1949. Mao, the leader of the peasant revolution, had to reconstruct the country, which had been devastated by more than 30 years of warfare.

firmly on Chinese society. They closed schools and universities and destroyed Buddhist and Taoist temples, hoping to eliminate all Western, intellectual, traditional, and religious influences so that the Chinese people would become purely Communist. The Cultural Revolution was bloody and violent, especially in its early years; many professors and others were sent to prison or to rural farms for "reeducation," which often meant brainwashing and forced labor. Throughout this period, Mao was head of the CCP and the supreme ruler of China; the post of premier was held by several men, the most powerful of whom was Zhou Enlai.

The United States and many other non-Communist nations refused to give diplomatic recognition to the People's Republic of China. After 1949, China's seat in the United Nations was occupied by Taiwan, which called itself the Republic of China (ROC). Several times during the 1960s, however, the PRC tested nuclear weapons, and the United States and other Western nations began to recognize that they could not continue to ignore the existence of a country that had one-fifth of the world's population as well as nuclear arms. In November 1971, a vote in the United Nations took China's seat away from the ROC and gave it to the PRC. The following year, Richard Nixon became the first U.S. president to visit China. He climbed onto the Great Wall, dined with Mao and Zhou, and generally paved the way for warmer relations between the United States and China. His visit was followed by diplomatic and cultural exchanges, trade agreements, and a surge of American tourists in China.

In 1975, Zhou Enlai announced that China would embark on a program called the Four Modernizations (of farming, industry, defense, and science) in order to become fully productive. Some of the new economic policies departed from strict socialist doctrine—for example, farm workers were to be permitted to cultivate private gardens and sell their produce in free markets after official quotas had been met. Factions sprang up in support of and opposition to these new policies. The opposition was led by Jiang Qing, Mao's wife, and three of her colleagues. This group, later called the Gang of Four, claimed that China must adhere strictly to socialist principles. A more moderate group, led by Deng Xiaoping (vice-premier and vice-chairman of the CCP), supported the Four Modernizations. After Zhou Enlai's death in January 1976, the Gang of Four gained power, and Deng was stripped of his official positions. The elderly Mao appointed Hua Guofeng as the new premier. Mao himself died in September 1976. Without his protection, the Gang of Four was arrested by Deng's supporters, and in early 1977, Deng

On May 16, 1989, Chinese leader Deng Xiaoping (center) shakes hands with Soviet leader Mikhail Gorbachev during a meeting at the Great Hall of the People in Beijing. Deng, who became China's leader after Mao's death, tried to rebuild China's economy by offering material incentives to workers and encouraging foreign investment.

reappeared as head of the CCP. For the next twenty years, Deng was the most powerful figure in Chinese politics. The Gang of Four stood trial in 1980–81, and all four members were sentenced to life imprisonment.

The cataclysm of the Cultural Revolution shook Chinese society, but in 1976 China was shaken in a different way. An earthquake that measured 8.2 on the Richter scale struck the city of Tangshan, about 90 miles (145 kilometers) east of Beijing. More than 650,000 people were killed in what is thought to have been the deadliest earthquake in recorded history. Houses and factories throughout the region were destroyed; rebuilding Tangshan was one of the government's highest priorities in the decade that followed.

Under Deng's leadership, major sections of the Chinese economy were freed from government control. Farming families acquired the right to own property and work for their own profit. Certain sections of the country, such as Shanghai, became "Special Enterprise Zones." In those areas, private citizens could start their own businesses and attempt to become rich. The new policies initiated a period of rapid economic growth. Between 1978 and 1997, agricultural workers saw their average incomes triple. The annual growth rate of the Chinese economy averaged 9 percent—a rate that would double the size of the economy every eight years.

The new economic freedom created a demand for political freedom. The most serious threat to Deng's regime came in June 1989, when thousands of students and workers in Beijing and elsewhere defied official orders and demonstrated for democratic reforms in the government. Deng's response to the protest was the massacre in Tiananmen Square, in which the People's Liberation Army dispersed the crowd and killed many demonstrators.

The violence in Tiananmen Square was followed by a period of repression. Many of the protesters were imprisoned. There was a slowdown in the economic reforms Deng had been supporting.

In 1992 Deng made a dramatic visit to southern China and again threw his support behind the push for a market-oriented economy. Younger, reform-minded leaders were given top positions at the 14th Party Congress later that year.

Deng died on February 19, 1997. Under his leadership, China had achieved dramatic economic progress. But it was still essentially a dictatorship ruled by a small committee. The struggle for democracy may disturb Deng's successors for most of the first decades of the new century.

A kindergarten student dances at a factory school in Beijing. The Han Chinese make up about 92 percent of the country's total population.

5

The Peoples of China

China is said to have a very homogenous population—that is, a population whose members are very similar, with the same culture, traditions, language, and ethnic background. In one sense this is true, because the overwhelming majority of China's people belong to the same broad ethnic category, the Han Chinese. The Han make up about 92 percent of the country's total population.

Although the population is dominated by the Han Chinese, three factors make China less homogenous than it appears. First, the population includes 55 non-Han ethnic groups, many of whose cultures and languages are quite distinct from those of the Han Chinese. Compared to the huge Han population, these ethnic groups are quite small, but most of them remain separate within Chinese society. Many of the non-Chinese people live in the five autonomous regions and the autonomous counties and townships, where they are permitted to practice and preserve their traditions. Second, the Han are concentrated in Manchuria and China proper, whereas the non-Chinese peoples are generally found close to border regions: in the southwest, north, west, and northwest. As a result, most of the Han Chinese live in about two-thirds of the country, and the remaining third of the land, although sparsely settled, has a greater proportion of non-Chinese to Chinese inhabitants. The Han are outnumbered by non-Chinese in the large autonomous regions of Tibet and Xinjiang Uygur. And third, even within China proper the Han population is not completely uniform. Over the centuries, the various districts and provinces developed their own customs, dialects, foods, and subcultures. Some of these regional and provincial traits remain very distinctive today, although China's population has grown more homogenous during the 20th century because massive relocation programs have moved large numbers of people around the country. Modernization plans and contemporary communications technology have also created a simplified form of the Chinese language that is used nationwide.

Overall, approximately 71 percent of the population lives in the countryside; about 29 percent lives in cities and towns. Since the 1950s, the government has relocated many millions of city dwellers to the countryside, hoping to reduce the pressure on the towns by making use of sparsely settled regions. At the same time, however, this city-to-country flow of people has been counteracted by a steady flow of rural peasants seeking jobs in the city. Today, many of the coastal provinces in the east contain over 400 people per square kilometer (1,000 per square mile). The most densely populated areas are the deltas of the Yangtze and Pearl rivers and the plains around Chengdu in Sichuan province. The North China Plain and the entire coastal belt are also heavily inhabited. By contrast, in the west and northwest—an area about as large as Europe—there are fewer than 50 people per square mile. As an even more dramatic example of these contrasts, Shanghai has a population density of 3,270 people per square mile (1,260 per square kilometer), whereas Tibet has only 5 people per square mile (about 2 per square kilometer).

The Han Chinese

The Han Chinese are the descendants of China's earliest known inhabitants in the Yellow River valley. They are typically somewhat shorter and slighter in stature than the average European, and almost without exception they have black hair and very dark brown eyes. Like some other East Asian peoples, the Han have what is called the epicanthic fold—a fold of skin on the upper eyelid that gives the appearance of narrowing the eye. Their skin color ranges from very fair to golden or olive brown.

As the Han people spread throughout China proper from the birthplace of their culture in the North China Plain, they carried with them their language, which existed in written form almost

A page from a 12th-century edition of a book illustrates the ancient musical instruments that were used for ceremonial purposes. Written Chinese does not use an alphabet but rather ideographs, or symbols that represent whole words, and many of these ideographs require as many as 25 strokes of the pen.

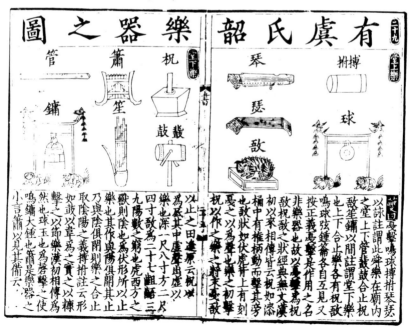

4,000 years ago. Over time, however, the spoken language took on different sounds in different parts of the country, while the written language remained the same. This resulted in many dialects, or regional varieties, of spoken Chinese but only one written language. Some of the dialects were so far removed from the original that people from one province could not talk with people from another province, but all educated Chinese could communicate in writing.

In modern times, Chinese dialects fall into two broad groups: northern and southeastern. The northern dialects are forms of Mandarin Chinese. In 1955, the Chinese government selected the version of Mandarin that is spoken in and around Beijing as the official national language, called *putonghua*, the "common language" or "ordinary language," and most Chinese in other parts of the country can now speak and understand at least some basic Mandarin. The dialects of the southeast, however, are often difficult and sometimes impossible for Mandarin speakers to understand. The most widespread of these dialects is that of Shanghai, sometimes called the Wu dialect. It is used in trade and business. Hakka and Hokkien are two dialects of the southeastern coast that are used by many of China's fishermen. The dialect of Guangdong province is called Cantonese; it is spoken by most of the Chinese who have immigrated to other countries.

Written Chinese does not use an alphabet, as do European languages. Instead, it uses ideographs, which are symbols that represent whole words. Traditionally, a writer of Chinese had to learn how to draw many thousands of ideographs—as many symbols as there were words. Many of these ideographs, or characters, as they are commonly called, are very elaborate. About 70 percent of the 2,000 most commonly used characters require 9 or more strokes of the pen, and many of them require as many as 25 strokes. It is easy to see why education remained the privilege of the wealthy and leisured classes for so many centuries.

Kuan Yin, the goddess of mercy, is a uniquely Chinese personification of Buddha. The statue in this photograph weighs more than 110 tons (111 metric tons) and measures 72 feet (22 meters) high.

Since 1949, as part of its plan to make education universal, the PRC government has been trying to simplify written Chinese. It has done so in two ways: by adopting simpler, easier-to-draw versions of the 2,000 to 3,000 most-used characters, and by adopting a spelling system called pinyin that uses the Latin alphabet (the same letters that are used in English) to write Chinese words. Pinyin is used in China's schools to help teach Chinese, especially to ethnic minorities, although students are still expected to learn the ideograph script as well. Pinyin is also used on signs and for place names, and most foreign governments and publications have adopted pinyin spelling for Chinese names.

Religion has also undergone changes in modern China. For centuries before the Communist revolution, the traditional religion of the Han Chinese was a mixture of Confucianism, Buddhism, and Taoism. Some Chinese considered themselves strict Buddhists for example, or strict Taoists, but most people shared a system of beliefs and rituals that drew upon all three religions or philosophies. Weddings, funerals, the Chinese New Year, and other festivals were generally celebrated with rituals and ceremonies from Taoism, Buddhism, and folk tradition, all at the same event. The Chinese believed in a large pantheon, a family of hundreds of gods, headed by the Supreme Ruler of Heaven. Buddha, Confucius, the Taoist spirits, and even Muhammad and Jesus Christ were often regarded as intermediaries between the Supreme Ruler and earth. The different religions were simply different ways of communicating with Heaven. Even the Buddha took on a new form in China; he was sometimes transformed into Kuan Yin, the Buddhist goddess of mercy, a uniquely Chinese personification of the Buddha. The most universal form of religious behavior was ancestor worship, in which the living showed reverence for dead ancestors by making offerings to family altars, tending the ancestors' graves, and performing rituals on days set aside to honor the ancestors. Ancestor worship was related to two important elements of Confucianism, family loyalty and submission to the father.

This ancient religious and ethical pattern has been greatly altered in the 20th century. In the early part of the century, many Chinese were converted to Christianity by Western missionaries. But the biggest change came after 1949, when the PRC discouraged religious practices, which were dismissed by the Communists as "feudal." This intolerance turned to outright persecution during the Cultural Revolution. Many temples, shrines, and churches were destroyed, and religion was regarded as an obstacle to progress and enlightenment. Since then, the government has eased its antireligious position, and people are no longer punished for taking part in religious observances. Today, approximately 100 million Chinese (about 8.3 percent of the population) are followers of

Taoism, Buddhism, Islam, Catholicism, and Protestantism. Many ethnic groups have their own religious preferences. For example, Tibetans, Mongolians, Lhobas, Moinbas, Tus, and Yugurs are generally Lamaists. The Hui, Uygur, Kazak, Kirgiz, Tatar, Ozbek, Tajik, Dongxiang, Salar, and Bonan follow Islam.

China's Ethnic Minorities

The Zhuang are China's largest minority group, making up 1.33 percent of the total population. They are culturally related to the people of Thailand. Some of them are Buddhists, and some practice a form of ancestor worship and spirit worship. The Zhuang have their own written and spoken language. They are concentrated in the Guangxi Zhuangzu autonomous region and the neighboring provinces of Yunnan and Guangdong, where they live largely by rice farming.

The Hui are Muslims, descendants of Chinese who adopted the religion of Islam when it entered China in the 7th century. The 8 million Hui make up 0.67 percent of the population. Most of them live in the Ningxia autonomous region and in smaller autonomous communities in the provinces of Gansu, Henan, and Hebei. The Hui use the Chinese language.

The Uygurs are also Muslims. They live in the Tarim and Junggar basins in the Xinjiang Uygur autonomous region. Their language belongs to the Turkic language family of Central Asia rather than to the Chinese language family. The Uygurs are traditionally tent dwellers and camel herders, but today many of them live in modern housing in settled oasis communities, where they have begun to practice gardening and small-scale farming. The Kazaks and the Kirghiz are two other Turkic Muslim peoples of western China. Both are nomadic herders, and both are related to ethnic groups that live in the neighboring states of Kazakhstan and Kyrgyzstan. Their numbers are small.

Farmers of Sichuan province spread out grains to dry in the summer sun. The Yi, an ethnic group that lives in Sichuan and Yunnan provinces, are mostly farmers and herders, and their language is related to Tibetan.

The Yi (also called the Lolo) live in Sichuan and Yunnan provinces. They are farmers and herders, and their language and customs are related to those of the Tibetans.

The Miao and Yao are ethnic groups of southern China. Their languages and cultures are a blend of Chinese and Tibetan influences, but both of these groups are rapidly being assimilated into the Han population. Less easily absorbed are the Wa, a small but distinctive population along the border of Myanmar. The Wa are related to the tribespeople of Myanmar. They live in small villages in the jungle-covered mountains and maintain a traditional way of life based on hunting.

Among the ethnic groups of northern China are the Manchu, who claim to be descended from the Manchurians who invaded China in the 17th century, and the Koreans, who live in an autonomous district in Jilin province, which borders North Korea.

Two of China's ethnic groups are small in number but large in importance because they are spread over huge areas of the country's strategically sensitive frontiers. These are the Tibetans (0.4 percent of the total population) and the Mongolians (0.4 percent). The Tibetans inhabit all of the Tibet autonomous region and much of Qinghai province; there are also Tibetan communities in western Sichuan and the Xinjiang region. Taller than the average Chinese, with features that resemble those of Native Americans, the Tibetans farm in the sheltered valleys of the plateau and graze herds of yaks and sheep on the plains. Some settlements in Tibet, including Lhasa, the capital, are quite old. The principal buildings throughout Tibet are Buddhist temples and monasteries because most Tibetans are Buddhists.

The Mongolians also practice Tibetan Buddhism. Most Chinese Mongolians live in the Inner Mongolia autonomous region, although they are also found in communities all the way from Xinjiang Uygur in the northwest to Jilin in the northeast. The Mongolians were nomadic herders for centuries, but today many of them live in settled towns and combine farming with herding. Among the Chinese they are still noted for their outstanding horsemanship—the skill that helped the Mongol hordes conquer most of Asia in the 12th and 13th centuries.

The PRC takes pride in the fact that it has allowed China's ethnic minorities to retain their cultural identities while at the same time improving their economic well-being, health, and education. Non-Chinese minorities are even encouraged to have larger families than the Han Chinese so that they will not be completely swallowed up by the Han population. Yet unrest is growing among some of the ethnic groups, particularly those of western China, where the non-Chinese outnumber the Chinese. Some non-Chinese feel that they are treated as second-class citizens and kept out of the mainstream of Chinese affairs, and others call for greater independence from the PRC. In recent years, the Tibetans have been most troublesome to the central government. Periodic uprisings against Han rule in Tibet have been crushed by force.

84

Aspects of Daily Life

Most rural settlements are compact villages surrounded by fields; in the north, the villages are connected to one another by cart tracks, footpaths, and, increasingly, roads. At the center of a ring of villages is a market town, which serves as the area's shopping and cultural hub. Most houses are built of sun-dried mud brick or cement blocks. In the south and west, where trees are more common than on the plains, some houses are of timber. Building materials are used over and over again, so that some houses in which families live today were rebuilt by their owners from timber and bricks used in their grandparents' house.

The southern landscape is dominated by rice fields, which are often flooded. Paths along raised dikes enable people to travel between fields. In the south, streams and canals link the villages, and most local travel is by boat. Not all rural Chinese live in villages; isolated homesteads are found amid the bamboo groves of the Sichuan plain and elsewhere in the south and west.

Mao Zedong encouraged the formation of communes, such as this one located near Beijing. Communes were divided into production brigades, made up of about 1,000 people, and their goal was to produce goods for their own use. Each commune had its own health clinic, stores, marketplace, and school.

Shoppers in Chengdu, the capital of Sichuan province, look over cuts of meat at an outdoor market. At the center of a circle of villages there is usually a market town, which serves as the region's shopping and cultural hub.

China suffers from an acute shortage of housing. The problem is worse in the cities, because they have grown rapidly during the 20th century, and construction in them has not kept pace with rising populations. Since 1949, urban construction has been one of the government's top social priorities. Such construction takes the form of large, square apartment blocks. The majority of city dwellers now live in these apartments, but many families share their quarters because there is still a shortage of apartment units.

In China before the Communist revolution, clothing was a clear sign of class distinctions: aristocrats wore elaborate embroidered silk ceremonial robes or sometimes, in the 20th century, Western-style clothing, whereas common people generally dressed in loose-fitting pajamalike cotton garments. In the years following 1949, all Chinese dressed in some version of the uniform that was favored by Mao Zedong. It consisted of a military-style jacket and trousers in dark blue or black cotton and a small cap. The Chinese people looked like a single giant army throughout the 1950s and 1960s.

Today custom permits a wider range of clothing, and more choices are available in the stores. China's leaders are usually seen

Top: *A woman from Inner Mongolia wears traditional clothing, which includes a long coatlike garment, called a del, that is often richly embroidered.* Bottom: *Many Mongolians live in yurts, circular domed tents of skins or felt stretched over a frame.*

in Western-style business suits. Most men in the cities wear Western-style trousers and shirts, and women wear dresses. Some Chinese even dress in miniskirts, jeans, and T-shirts, especially in the more sophisticated coastal cities, such as Shanghai and Guangzhou. Many rural Chinese still dress in military-style work uniforms, although in the south some farm workers have returned to their traditional garb of woven straw hats, loose jackets, and long strips of cotton worn as wraparound skirts or loincloths. China's ethnic populations are free to dress in their traditional clothing if they wish to do so. The Tibetans, Uygurs, and Mongolians have the most distinctive traditional clothing, with thick sheepskin caps and robes, baggy pants, and in the case of the Tibetans, much jewelry made of silver, turquoise, and coral.

One aspect of Chinese life that is familiar to people around the world is food. Emigrant Chinese have carried their native cooking methods to many countries. Although there are many different types of traditional Chinese cooking, most of them share the same basic principle: Meat or fish and vegetables are chopped or sliced into small pieces and then mixed together to be quickly fried in oil or cooked in some form of stew. Like many East Asian peoples, the Chinese use two slender sticks, called chopsticks, as eating utensils.

The ingredients used in Chinese cooking vary greatly from region to region, depending upon what products are locally available. This has given many provinces traditional, distinctive cuisines. Sichuan food, for example, is hotter and spicier than Cantonese food because Sichuan cooks use more chili peppers. Garlic, soy sauce, mustard, and ginger give pronounced flavors to many dishes. Wheat noodles are the staple carbohydrate of northern Chinese meals, but their place is taken by rice in the southern half of the country. The people of the far north and west, where animal herding has been the basis of the economy for centuries, eat much more meat and milk than people in other regions. Tea and beer are the most popular beverages.

A one-child family walks near the Leap Forward Commune in Xiling Hot, Inner Mongolia. The Chinese government has tried to curb the country's population growth by giving cash bonuses and better housing to those couples who have only one child.

6

Government and Social Conditions

The dominant force in the government of the People's Republic of China is the Chinese Communist party (CCP). It has been the only significant political party in China since 1949. Although only 57 million Chinese are members of the CCP, the party controls all government functions, the armed forces, and the schools.

According to China's constitution, the highest-ranking state body is the National People's Congress (NPC), a legislature of 3,000 delegates that meets in Beijing for two weeks each year to pass new laws. Members of local-level people's congresses are elected by the people—all Chinese over 18 years old can vote, unless they have had their voting privileges taken away by the state. The delegates to the local people's congresses elect delegates to district congresses, who in turn elect delegates to provincial congresses, and so on to the NPC, which is supposed to be the voice of the people. In reality, however, the NPC has little power. Most of the time it merely gives formal agreement to laws and plans developed by the CCP.

The highest-ranking government executive is the premier, who is appointed by the leaders of the CCP. The premier and the vice-premiers form a body called the State Council, which is significantly more powerful than the NPC. The State Council shapes China's laws, economic plans, and overall policies under the guidance of the CCP. It is also responsible for most of the decision making and general administration involved in the day-to-day operation of the government. The 1982 constitution restored the position of president, which had been eliminated by Mao Zedong. Despite being the formal head of state, the president has little real importance.

Local government is carried out by a huge bureaucracy that is shaped like a pyramid. The central government, which is relatively small and consists of the State Council and the heads of the CCP, is at the top. Taking its orders from the central government is a spreading network of regional, provincial, district, county, township, and community government bureaus. In all, about 10 million people work in the Chinese government—more than the population of some small countries.

Like the central government, all of the lower-level government bureaus are guided and supervised by the CCP, which also has a pyramid-shaped organization that reaches into every community. Officially, the top position in the CCP is that of secretary, but the real power in the party belongs to the person who is chairman of the party's four-member Central Advisory Commission. The National Party Congress, a meeting of high-ranking CCP members, takes place every five years and selects members to serve on the CCP Central Committee and Political Bureau, which are supposed to determine party policy. In reality, these bodies are directed by the Central Advisory Commission.

When the Chinese Communists founded the PRC in 1949, they modeled its judicial system on that of the Soviet Union. Since that time, the CCP has often ignored formal legal procedures. People who are suspected of disagreeing with the central government, for example, are often imprisoned without trial or tried without legal protection. The party's disregard for human and legal rights reached a peak during the Cultural Revolution, when many thousands of people were jailed by the Red Guards or sent to the countryside to be reformed by manual labor. During the late 1970s and the 1980s, the government claimed that it was trying to make the legal system more consistent, but party leaders continue to override the law at will. Amnesty International, a human rights organization, claims that 10,000 to 30,000 dissidents were arrested in the wake of the 1989 protests alone.

Since the early 1980s, lawyers have been permitted to practice on behalf of private clients, but so far China has only a few thousand of them. The law allows local CCP committees to sentence "hoodlums" or "antisocial elements" to as much as four years in a labor camp (generally located in western China), with no opportunity for appeal. Beginning in 1983, the government launched a strict anticrime campaign, and for a while public executions—which are the punishment for about 25 different crimes—occurred at the rate of

The Chinese People's Liberation Army marches in a parade. In the late 1990s, China had about 3 million troops.

In an effort to improve his hearing, a 12-year-old patient undergoes acupuncture treatment at the Lo Ping School for the Deaf in Shanghai. Traditional medicine uses acupuncture and herbal remedies.

about 200 a month. A government bureau called the Ministry of Justice oversees the judicial system. There are four levels of courts, ranging from Basic People's Courts to the Supreme People's Court.

The Chinese military consists of the People's Liberation Army, which includes land, sea, and air forces. All men must perform military service after age 18; women can enlist for medical or technical duty. In 1997, the People's Liberation Army had about 2.9 million members, making it the world's largest armed force. Police work is carried out by the Public Security Force, which takes its orders from the CCP.

In China, many social services—such as medical benefits and pensions—are provided by large unions, professional and trade organizations, factories, and communes (political units to which nearly all rural households belong). Both law and custom require working people to take care of their unemployed and elderly relatives, usually in their own home.

Health care has improved dramatically during the 20th century. Diseases such as smallpox, cholera, and typhus that once ravaged the population with frequent epidemics have been brought under control through massive immunization programs in the 1950s and 1960s. Malaria and schistosomiasis (a parasitic infection) were once widespread in the south but have been greatly reduced. The life expectancy of the average Chinese has increased by about 30 years since 1949, to nearly 71 years for women and 68 for men. The most frequent causes of death in China today are cancer, heart disease, and respiratory diseases such as emphysema; the severely polluted air that surrounds most industrial centers is thought to contribute to lung disease.

Medicine is practiced in two forms in China: traditional and Western. Traditional medicine uses acupuncture (tiny needles inserted into the skin at special points on the body to relieve pain or cure disease) and herbal remedies; Western medicine uses modern drugs and surgical procedures. Doctors of Western medicine outnumber traditional healers, especially in the cities, where most of China's hospitals are located. But many of the more remote rural

On March 12, 1979, Chairman Hua Guofang (waving), Vice-chairman Deng Xiaoping (with boy), and other Chinese Communist party leaders join more than 1,000 people in planting trees outside Beijing. March 12 is China's national tree-planting day.

In June 1989, students at the University of Beijing protest the government's offer of money and a loaf of bread to people willing to march in a progovernment rally. After the anti-government protests in 1989, the CCP abolished many student clubs and organizations.

districts do not have adequate medical care of any kind, and the country has only 1 doctor for every 633 people and 1 hospital bed for every 427 people. After the Communist revolution, the PRC sent medical students and assistants into the countryside to provide basic health care. These "barefoot doctors," most of whom receive about two months of training, are still an important part of the country's health care program.

Birth control is part of medical care and social planning. In an attempt to slow the growth rate of China's immense population, PRC leaders have passed laws aimed at reducing family size. Abortion is legal, and sterilization and birth control materials are free. Couples who have only one child receive privileges such as cash bonuses and better housing, whereas couples who have more more than two children suffer penalties such as fines and reduced food rations. Most urban families are small, but the old tradition of large families has remained alive in the country, where children take part in agricultural labor.

The CCP is proud of its record in improving literacy in China. Before 1949, 85 percent of all people over 15 years of age were illiterate—that is, they could neither read nor write. A huge literacy program during the 1950s and 1960s brought basic language skills to millions of peasants. Today only 6 percent of the people between 15 and 47 are illiterate. The educational system provides six years of primary school followed by six years of secondary school. Nearly all children enter primary school, but only 65 percent finish it; the dropout rate is higher in the country than in the cities. In general, urban schools are bigger and better-equipped than rural ones. Students who complete primary school must pass examinations to go to secondary school, and over 86 percent of them do so, according to official government figures.

China has over one thousand colleges and universities. Approximately 3 million students are enrolled in institutions of higher learning. Each province has a provincial university, but the country's biggest and best universities are the University of Beijing, Qinghua University in Beijing, the University of Nanjing, and Tianjin University. There are also a number of technical schools and agricultural colleges.

For most of the history of the People's Republic, students attended college free of charge and the government assigned them a job when they graduated. The government told them what they would do and where they would work. In 1994, some colleges switched to a different system. Students paid for part of their education and found jobs on their own when they graduated. In the fall of 1997, every college and university in the country made the change, and students started paying $180 per year in tuition. Students who come from poor families receive loans, subsidies, and jobs in work-study programs.

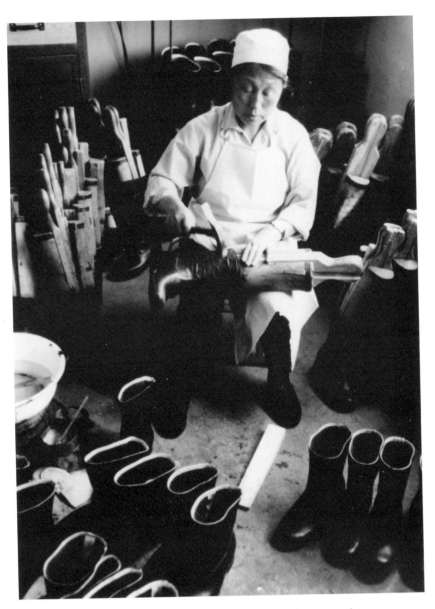

A factory worker makes boots in Hohhot, Inner Mongolia. Urban women have many more career choices than do country women, and women who work in factories earn a wage equal to that of men.

7

Economy

The size of China's population has long made simply feeding and clothing the people a major economic enterprise. Agriculture continues to be the mainstay of the economy. Because much of the country is covered with mountains, deserts, or poor soil, only about 10 percent of the land is suitable for farming. Traditionally, China produced just enough food to sustain itself each year, barring droughts or floods. During the three decades before 1949, however, the country's farmlands and farmers were devastated by wide-ranging civil war, the Japanese invasion, and the lack of any centralized economic planning. As a result, when the Communists finally took over, the country's economic output had dropped to a new low and hundreds of thousands of people were starving. The primary economic goal of the PRC government was thus to increase agricultural production and make China self-sufficient.

The large farms and estates of the wealthy were broken up and distributed to the peasants. The nation's farms were then organized into political and economic units called people's communes, which functioned somewhat like rural townships. They were responsible for local government, for meeting production levels set by the central government, and for carrying out projects like building schools and maintaining irrigation canals.

For many years the communes were an ideological focus for the Communist party. But the market-oriented reforms that were instituted in 1979 transformed the commune system. Rural Chinese were given the chance to work for profit, with less interference from the central government. By the late 1990s, they had formed over 18 million "Township and Village Enterprises," which account for 40 percent of China's industrial production and employ over 113 million people. They are active in industry, agriculture, transportation, communications, and other services. The goods they produce include food, household necessities, materials for light industry, and agricultural products.

Grains make up 76 percent of China's farm production, with the principal varieties being rice (more than 90 percent of all the rice produced in the world is grown in China), wheat, sorghum, corn, and millet. Peanuts, soybeans, potatoes, vegetables, and fruits are also cultivated. Crops for industrial uses include cotton (the basis for China's large textile industry), sugar, tobacco, tea, and rubber. Silk is another important industrial crop; it is made by weaving strands of a fiber made by silkworms, which feed upon the leaves of mulberry trees. Some attempt has been made to mechanize this age-old industry, but much of the work must still be done by hand. Livestock include pigs (China has more pigs than any country in the world—about 400 million of them are raised each year), sheep, goats (used increasingly for milk production), cattle, water buffalo (in the south), horses (mostly in the north), and camels (in Inner Mongolia and Xinjiang).

The fishing industry is concentrated in the south and southeast, in Guangdong, Fujian, and Zhejiang provinces. Many types of fish are caught, and clams, scallops, shrimp, and edible seaweed are cultivated on marine "farms." The country also possesses a network of 3,000 freshwater fish farms. Each year China produces

Women mend fishing nets on the banks of the Min River near Lushan in Sichuan province. Although the fishing industry is primarily concentrated in the southeastern provinces, China has a large network of lakes and ponds where carp and catfish are produced on fish farms.

9 million tons of freshwater fish—almost half the world total. Carp and catfish are among the favorites.

Forestry is carried out in Manchuria, northern Inner Mongolia, Sichuan, and Yunnan. China's forests are disappearing rapidly, and the government has started planting programs to reforest some areas, but neglect of the new seedlings and the practice of using them for fuel has hurt the effectiveness of these programs. In recent years, rice straw and other nonwood materials have been substituted for wood pulp in the manufacture of paper.

The country is rich in mineral resources: Geological surveys have shown that it has large deposits of tin, gold, iron, coal, lead, zinc, and manganese as well as some natural gas, but the government does not yet have the money, skills, or equipment to exploit these resources fully. Most mining is for coal, iron ore, tin, and tungsten. Nearly all provinces have coal deposits, and coal remains the country's principal fuel for households and electrical plants, although mining and burning it have caused serious environmental

Miners remove coal from one of China's many mines. Nearly all provinces have coal deposits, and coal is the country's principal fuel for households and electrical plants.

problems. China also has several petroleum fields in Manchuria and in Bo Hai. The oil is used to fuel industry. Energy is also generated by water power. The Gezhouba Dam on the lower Yangtze River is China's largest hydroelectric project.

China's leaders have been determined to make their country a modern industrial nation, and government-owned factories sprang up everywhere during the 1960s and 1970s. Most industry is concentrated around the major coastal cities such as Shanghai and Tianjin, but many factories have also been built at inland locations in an attempt to spread industry and population more evenly. As a result, such cities as Baotou in Inner Mongolia, Chongqing in Sichuan, and Harbin in Manchuria have become important centers of production and gained huge populations in a short time. Industry is on the rise in Beijing, Xi'an, and Luoyang. Anshan, in Liaoning province, has become the center of the steel industry.

Today about 16 percent of the labor force works in manufacturing industries. The leading industrial products include iron and steel, machinery, armaments, textiles and clothing, cement, chemical fertilizers, autos, consumer electronics, and telecommunications equipment.

The tourism industry has also been important for China. Tourism flourished after President Nixon's historic visit in 1972, as the non-Communist world rediscovered a country that had been off-limits for decades. Tourism dropped off sharply in 1989 and 1990 in the wake of the Tiananmen Square violence, but it soon surpassed its earlier levels. By the last years of the twentieth century, tourists from around the world—many of them people of Chinese descent—were bringing over 8 billion U.S. dollars into China each year.

In the early years of the Communist regime, China limited its foreign trade to the Soviet Union and the Communist nations of Eastern Europe. After the split between China and the Soviet Union in the 1960s, China developed trade relationships with non-Communist nations, although it continued to trade with

China's modernization efforts have included computerization in key industries and government departments. This early computer center in Beijing, installed by the United Nations Fund for Population Activities, processed the 1982 national census. Since then, China has developed a growing computer industry of its own, with significant production of consumer electronics goods and telecommunications equipment.

the Soviets. Today its principal trading partners are Japan, the United States, Germany, South Korea, Singapore, and Taiwan. China's major exports are garments, textiles, footwear, toys, and machinery. Its major imports are industrial machinery, textiles, plastics, telecommunications equipment, steel bars, and aircraft.

The joining of Hong Kong to China in mid-1997 created both major opportunities and major issues for China. Since Hong Kong had substantial trade and investment ties with Western nations—for example, there were about 900,000 U.S. firms operating in Hong Kong before the merger with China—the Chinese stood to benefit enormously. However, some analysts worry that if China interferes too greatly in the Hong Kong economy, or restricts the liberties of Hong Kong citizens too obviously, Western firms may withdraw some of their investment.

The currency used in China is called renminbi, or "people's money." Its basic unit is the yuan, which is equal to about 12 U.S. cents. By the mid-1990s, the average annual income in rural areas had reached U.S. $230 per person; city dwellers averaged about U.S. $525 per person.

Transportation

Ever since the Tang emperors first began to build the Grand Canal—more than 2,000 years ago—waterways have been important avenues of transportation. Wooden junks, built to a centuries-old design, are still common on the inland waterways. Shanghai, Guangzhou, Hong Kong, and Tianjin are the major ocean ports.

Railways are the backbone of the transportation system. China has about 34,000 miles (55,000 kilometers) of track, and for many years most goods and passengers have traveled by rail. Highways, however, are becoming increasingly important. China has about 4 million cars and buses and about 6 million

The Yangtze River Bridge at Nanjing is about 3 miles (5 kilometers) long and was completed in 1968. More than 50,000 construction workers built the bridge, which has 2 tiers: the bottom level accommodates railroad traffic, whereas the upper level is for regular vehicular traffic.

trucks. By the mid-1990s, China's highways were carrying 9 billion tons of freight and 9 billion passengers per year.

Between 1970 and 1990, major roads were built between Tibet and Sichuan, Xinjiang Uygur, and Qinghai. The major highways under construction in the late 1990s included Shanghai–Nanjing, Shanghai–Hangzhou, Hangzhou–Ningbo, and Guangzhou–Zhuhai.

China has international airports at Beijing, Shanghai, Guangzhou, Hong Kong, Tianjin, Ürümqi, Nanning, Kunming, and Hangzhou, as well as many smaller airfields. The government operates a domestic airline that offers daily flights between most of the major cities.

A group of people do their morning exercises in a Beijing park. At many factories and workplaces, workers spend their morning breaks exercising to music. After work, people play basketball, table tennis, and volleyball, which is China's most popular sport.

8

Culture and the Arts

For years radio was the dominant form of mass media in China. By the late 1990s, however, official government sources were claiming that most Chinese households owned televisions. In urban areas, according to these statistics, there were 116 TVs and 76 VCRs per 100 households. In rural areas, there were 76 TVs per 100 households. In the late 1990s, China still possessed only one telephone for every 60 persons, but telephone lines were being expanded. Cellular phone systems and a domestic satellite system had been set up.

The government controls all radio and TV broadcasts. The state-run Chinese Central Television (CCTV) has five channels that broadcast in Chinese. In rural districts, the government also operates a massive public broadcast system to communicate news, propaganda, and music to the masses; this system is wired to millions of loudspeakers and is the only one of its kind in the world.

The country has 1,755 newspapers and about 7,000 magazines and periodicals, all closely monitored by the CCP. The weekly *Beijing Review* is distributed abroad in several foreign-language editions, and the *China Daily News* is published in English. The largest newspaper is the *People's Daily*, which has a circulation of 2.15 million.

Some popular sports are uniquely Chinese. These include forms of martial arts such as wushu (known in the West as kung fu) and tai chi chuan, or tai chi for short. Wrestling is another traditional Chinese and Mongolian sport. In modern times, Western sports such as ping pong, soccer, tennis, weight lifting, gymnastics, and basketball have gained popularity. Among the ethnic peoples, the Mongolians practice horse racing and archery, the Tibetans race horses and yaks, and the Uygurs have a traditional horseback sport called the sheep chase, in which rival teams battle for possession of a sheep's carcass.

Chinese society contains many clubs and organizations, some of which have tens of millions of members. Some, such as the Communist Youth League and the Young Pioneers, are dedicated to furthering Communist goals. Others, such as the Federation of Literary and Art Circles, the Women's Federation, and the All-China Federation of Trade Unions, serve social or professional needs. All mass organizations, however, are closely supervised by the CCP, which can shut them down for any antigovernment activity. Student groups have fallen under particularly close scrutiny since 1989.

One of the leading cultural institutions is the National Library in Beijing, the country's largest library, with more than 10 million volumes, including many ancient scrolls. The Chinese Academy of Sciences Central Library, also located in Beijing, has branches in four other cities. Beijing's most splendid cultural monument, however, is the Forbidden City, a walled city-within-a-city of palaces and temples that formerly belonged to the imperial family and is now maintained as a museum. Part of it, the Imperial Palace Museum, houses a collection of rare paintings, sculpture, furnishings, and silks. Other notable museums are Beijing's Museum of the Chinese Revolution, which covers history from the 1850s to the present, and the Museum of Art and History in Shanghai, which has many exhibits on China's archaeology and early history.

The Arts Past and Present

China has one of the world's richest and most diverse artistic heritages. The oldest Chinese arts are probably music and dance. One of China's earliest archaeological relics, a 5,000-year-old pottery bowl from Qinghai province, is decorated with images of musicians and dancers. Bells, drums, gongs, flutes, and other instruments have been found in many ancient tombs. The next art form to develop was the bronze casting of objects such as vases and jewelry. Literature was flourishing by the time of the Eastern Zhou dynasty in the 6th and 5th centuries B.C. Many of China's classic works of philosophy date from this period, as does one of its earliest literary works, a collection of folk songs and ritual hymns called the *Shih ching*. China's first known paintings were on cave walls, but by the time of the Han dynasty (206 B.C.–A.D. 220) painters were at work producing landscapes on scrolls. China's distinctive architecture had emerged by that time as well. The traditional styles featured pagoda roofs, which are relatively low and flat and swept up at the corners, and square buildings surrounded by pillars. Larger buildings were created by adding more square, pillared units, often around a courtyard.

From Han dynasty times on, practitioners of the arts fell into two classes: One class consisted of skilled craftsmen, usually anonymous, who produced wares for sale or at the command of the court. The other class consisted of leisured, wealthy individuals, usually nobles, who developed literary or artistic tastes and produced works for their own pleasure. Their productions were highly prized by their friends and contemporaries, and many of them have been preserved. Court culture and popular culture developed distinct styles. Court artists produced elegant essays and poems, novels about aristocratic or political life, and formal music; while popular artists engaged in folk tales and melodies, juggling and acrobatics, and comical plays with such standard characters as unfaithful wives, blustering generals, and dishonest merchants.

A Beijing opera company performs "Beijing Opera," a kind of Chinese opera that uses a classical tale to portray the struggle between good and evil. This style of opera is also characterized by dancing, elaborate costumes, and displays of martial arts and acrobatics.

Court and folk traditions often met and merged, as in the Chinese opera, which has long been popular with all levels of society. There are about 300 different varieties of opera in China, but the best known of them is called Beijing opera. This style of opera is characterized by long programs of 40 or 50 scenes based on historical themes, with singing and dancing, elaborate masks and costumes that have traditional meanings, and displays of martial arts and acrobatics. Some Beijing operas include magical or spiritual elements drawn from Taoism. It is not uncommon for a performance to feature only the best-known scenes from several operas.

A new element entered artistic life during the 20th century. The rise to power of the Communist party brought both new subject matter and new limitations to artists and writers. The CCP holds the view expressed by Mao Zedong in the 1940s that art must serve the needs of the party. After 1949, artists were called upon to paint scenes of social realism, such as factories being built or peasants working in the fields, which were completely divorced from Chinese artistic tradition, and writers were expected to produce works that glorified the heroes and goals of the revolution. New operas were written about the revolution; their songs bore such titles as "The Red Detachment of Women," "Looking Forward to the Liberation of the Working People of the World," and "Socialism is Good." Since the 1970s, however, artists and writers have been given greater freedom to use traditional styles and nonpolitical subjects. But freedom of artistic expression is far from complete, and no works that criticize the government are permitted.

China's literary tradition embraces history, philosophy, poetry, essays, and fiction. Examples of all of these types of writing survive from more than 2,000 years ago. K'ung Fu-tzu and Lao-tzu and their followers began a centuries-long process of philosophical discussion and debate, and Ssu-ma Ch'ien set the pattern for historical writers. One of China's earliest literary figures was the poet Chu Yuan (340–278 B.C.), who wrote melancholy love poems; the best known of his works is called "Falling into Trouble." A later poet, T'ao Ch'ien (A.D. 365–427), spent his entire life on a small farm, drinking wine and writing *yueh-fu* (personal, lyrical poems in ballad form) that celebrated the beauties of nature and the changing of the seasons. He is regarded as one of China's greatest writers.

The Tang dynasty is sometimes called the golden age of Chinese poetry. Two poets of the period, Li Po and Tu Fu, were masters of the *lu-shih*, a poem in eight lines with elaborate rhyme schemes. Many of Li Po's poems were odes to beautiful landscapes, whereas Tu Fu described the hardships and simple lives of ordinary people. Because the two men were friends, some of their poems are messages to each other, concerned with the themes of joyful meetings and sorrowful partings. All of these themes became standard in the works of later poets. Po Chu-yi, a poet of the later Tang dynasty (772–846), wrote gentle ballads that became universally popular in China and have been translated into many languages. Two important poets of the Song dynasty were Su Tung-po, who wrote poetic essays about landscapes, and Lu Fu, who wrote about 20,000 poems, many of which were patriotic anthems urging the Song to defend themselves against the advancing Mongols. In the 17th century, the Manchu poet Na-lan Qing-de wrote works inspired by the natural beauties and vast open plains of his native Manchuria. In the 20th century, writers such as Huang Tsun-hsien have tried to revitalize poetry by abandoning old styles and subjects in favor of everyday language and realistic subjects. Pei Tao is one contemporary poet who experiments with new poetic styles and symbols. Contemporary playwrights, drawing upon the long-standing love of the Chinese people for the theater, have found audiences for both propaganda plays and nonpolitical ones. One very popular modern play is *The White-Haired Girl* written in 1953 by Ho Ching-chih, which is based upon folk legends.

Fiction got its start with folk tales and historical anecdotes that were embroidered by writers. One early fictional form was the *chuan-chi*, or "tale of strange things," a short story, usually a love story, with elements of magic or the supernatural. The novel emerged during the Yuan period, but the early novels were not the works of specific authors, rather they were popular stories, generally lengthy historical epics, set down in various form by many writers. One of the earliest of these was *The Romance of the Three Kingdoms*; the version of this novel that is most widely read today probably dates from the 16th century.

During the Qing dynasty, Ts'ao Hsueh-ch'in (also called Ts'ao Chan) wrote a novel that is considered to be a masterpiece of world literature. Called *The Dream of the Red Chamber*, it is a tragic love story about the fall of a powerful family. Among the leading novelists of the modern era are Lu Xun, or Lusin (1881–1936), who satirized Chinese society in the early 20th century; Ba Jin, whose

Chinese workers take a break at a Beijing factory. Behind them, a painting depicts various scenes of progress made by the Communist party, such as the military, satellite technology, and aviation. The CCP holds the view that art must serve the needs of the party.

autobiographical novel *The Family* (1931) depicts the conflict between traditional society and youthful revolutionaries; Mao Dun, whose 1933 novel *Midnight* explores the corruption and chaos of Shanghai on the eve of the Japanese invasion; and Lao Tze, a comic writer who turned his hand to social realism and produced a moving novel about poverty in *Rickshaw Boy* (1936). After the revolution, many novelists concerned themselves with social issues, such as the female writer Ding Ling's exploration of land reform in *The Sun Shines over the Sanggan River* (1949). In more recent years, short-story writers such as Wang Meng (former minister of culture) have experimented with Western styles and a broader range of subjects.

The visual arts exist in two broad categories in China: painting, which traditionally depicts landscapes or standard scenes such as greetings between scholars; and calligraphy, or fine ornamental writing. The same tools—a brush, sooty black ink, and silk or rice paper—are used for both; sometimes watercolors are also used in painting. Chinese landscape painting reached its peak in the Song and Ming dynasties. It was thought that the artist's sensitive portrayal of rocks, water, and plants, often without people or signs of human habitation, gave both the painter and the viewer an opportunity to meditate on the forces that hold the natural world together—light and darkness, wind and water.

The works of renowned painters from many dynasties have been preserved. It was common for painters to copy the works of earlier masters. Thus a landscape by the 14th-century artist Ni Tsan was reproduced by the 17th-century painter Wang Yuan-ch'i. For this reason it is sometimes impossible to determine how much of a particular artist's style was original. For the most part, faithfulness to traditional styles and excellence of execution have been valued above originality in Chinese painting.

Traditional Chinese art contains almost no scenes of violence, nudity, or death. It is rich in symbols drawn from nature: the crane,

In June 1989, 14-year-old artist Wang Yani worked on a painting on the floor of the Smithsonian Institute's Sackler Gallery in Washington, D.C. Called "a very rare phenomenon anywhere in the world," Wang became very popular in her native China.

for example, represents long life; bamboo represents the scholar; and gnarled and twisted pine trees represent old age. Buddhism gave artists another realm of subject matter, and the various forms and adventures of the Buddha are portrayed in statues, scrolls, and cave paintings. Western China has many fine examples of Buddhist art; perhaps the most celebrated is the Cave of a Thousand Buddhas at Dunhuang in western Gansu province.

In more modern times, Ch'i Pai-shi (1863–1957) and other painters have combined traditional Chinese subjects with Western materials such as oil paints and Western styles such as Impressionism. For some time after the Communist revolution, artists were limited to politically acceptable subjects, and most new art took the form of murals glorifying socialism and the revolution. The recent return to more traditional subjects and methods is illustrated in the paintings of Wang Yani, a young woman whose lively ink drawings of animals and landscapes became very popular in China while she was still a teenager.

China is also noted for its distinctive art objects made by craftspeople over the years. Among the characteristic art forms are cloisonné (brightly colored enamel designs on metal), carved jade, bronze statues (often of the Buddha or of animals with mythological importance such as dragons and tigers), and fine porcelain vases and bowls. So much pottery was exported from China during the

A wooden scroll box from the Yuan dynasty (1279–1368) is an example of Chinese lacquerwork. After many layers of lacquer have been applied to the object, intricate designs are carved into the wood, exposing the various shades of the hardened lacquer.

early years of East-West trade that the term *china* came into general use for all porcelain ware. Among the best-known types of Chinese pottery are celadon ware, which has a fine greenish luster and was developed during the Song dynasty, and blue-and-white ware, which became popular during the Ming dynasty. Much blue-and-white ware was manufactured for export, and many pieces were painted with such scenes as the "willow pattern" landscape, which was copied by pottery makers around the world and is still in wide use.

Silk making was perfected in China thousands of years ago. Although the art of silk manufacture later spread to other parts of the world, Chinese silk has always been regarded as the finest. Tapestries and ornamental hangings, bolts of embroidered silk, and fine white silk used for painting have been preserved from all periods of China's history. Many patterns found in ancient silk artworks are reproduced on factory-made textiles today.

Examples of the Chinese arts can be found in museums and private collections all over the world. The single largest collection of masterpieces is in the Palace Museum of Taiwan; many art objects were brought to the island by Nationalists fleeing the Communist takeover in the late 1940s.

In 1927, a lamplighter makes his rounds in a village. Even today, China is a mixture of old traditions and new. It has struggled to achieve a balance between ancient traditions and newer values.

9

China in Review

The People's Republic of China occupies a prominent place in world affairs for three reasons: its large population, its great size, and its historical importance. In terms of population, it is the world's largest country, with more than 1.2 billion people, or 20 percent of the planet's total population. In terms of size, it is the third largest country in the world and dominates East Asia. And in terms of its history, for more than 2,000 years, the Chinese empire developed its own sophisticated, highly advanced culture—a culture that motivated explorers, travelers, and merchants in other parts of the world to open new land and sea paths to the East in order to sample China's wealth and glory.

Although the imperial courts maintained an elegant, cultivated way of life, the masses of the Chinese peasantry, for the most part uneducated and condemned to backbreaking agricultural labor, led lives that amounted to slavery. In the 19th and 20th centuries, when the imperial house had grown weak because of internal bickering and domination by foreign powers, new social movements— democracy and socialism—began stirring in China. The first half of the 20th century can be viewed as one long battle for the soul of the

118

*In 1997, Hong Kong re-
verted to Chinese control.
How strongly this change
will affect capitalism in
Hong Kong (and its stock
exchange shown here) is
uncertain. Analysts are
also observing the reper-
cussions on the internal
struggle in China between
the traditional socialist
factions and those wanting
economic and democratic
reforms.*

country, in which generals, regional warlords, members of the fading royal family, Western merchants, Japanese invaders, Communists, and Nationalists strove to carve out and enlarge their domains. The battle ended in 1949 with communism victorious. China became one of the two powerful forces in international communism (the other was the Soviet Union).

Since that time, the leaders and people of China have struggled to repair the damage done by decades of war, to build an efficient economy, and to achieve a balance between ancient traditions and more modern values. Economic progress has been uneven, and the old and the new often coexist uneasily. China was estranged from the United States and other non-Communist nations for several decades after the revolution, but in the 1970s the PRC government adopted an "open-door" policy that opened China to Western trade, tourism, and cultural influences. One result of looser restrictions in China was the growth of a movement that opposed the government. In the late 1980s, that opposition movement took shape as the China Democracy Movement, a coalition of groups—mostly students—that wanted democratic reforms in the government and economy.

The China Democracy Movement and the Chinese Communist party faced off in June 1989—at the same time that Mikhail Gorbachev was launching political and financial reforms in the Soviet Union and the Communist governments of Eastern Europe were being overthrown by popular uprisings. When it ordered the military to attack the prodemocracy demonstrators, the CCP told the Chinese opposition movement and the world that no democratic reforms could be expected in China.

Deng Xiaoping and his followers believed that Gorbachev had made a mistake when he tried to combine an increase in economic freedom with an increase in political freedom. They argued that economic progress had to come first. But how long can a central committee control people who are beginning to enjoy the benefits of the modern world? Deng is now gone, and the Chinese people are liberating themselves from centuries of poverty. Some observers believe that within a few years the Chinese government will be forced to expand social and political rights to match its citizens' rising economic power.

A Note on Chinese Pronunciation

The process of making a language such as Chinese pronounce-
able for Westerners is called *romanization* because most Western
alphabets are based on letters devised by the ancient Romans.
Over time, several different systems of romanization have been
used for Chinese and spellings are inconsistent. For the most
part, this book uses the modern *pinyin* system adopted by the
Chinese themselves, but certain historical names have come to be
accepted with their older spellings, and where appropriate the
older system of romanization has been retained. Before *pinyin*
there was the *Yale* system, and before the *Yale* system there was
the *Wade-Giles* system.

A double ss or hs or ih combination, or an apostrophe (which has
no practical effect on pronunciation), is a clear giveaway to the
use of the older Wade-Giles system. The use of c versus ts will dis-
tinguish pinyin from Yale. In general, the Yale system gives the
American speaker the most correct sounds. For example, in
pinyin the resort city of Kweilin is pronounced Guilin, and the
city of Kwangchow is Guangzhou.

The following letters are used in all systems and are pronounced
more or less as in English: a (a as in *ah*, not as in *ate*), b, d, f, h, k,
l, m, n, o (as in *boat*; except ong = u as in *sung*), p, sh, s, t, u (u as in
suit; except iu = o as in *boat*), w, y.

Pinyin	Yale	Wade-Giles	Pronunciation
ai	ai	ai	i as in *kite*
ao	au	ao	o as in *cow*
b	b	p	b or p as in English
c	ts	ts', tz'	ts as in *tse-tse* fly
ch	ch	ch'	ch as in *church*
d	d	t	d or t as in English
e	e	e, eh	e as in *fence*; except *eng* = *ung*
ei	ei	ei	a as in *bay*
er	er	erh	ar as in *car*
g	g	k	g or k as in English
i	i, r	i, ih	i as in *kite*; except i = rr as in *burr* after s, sh, ch, z, zh
q	ch	ch'	ch as in *church*
s	s	s, ss	s as in *sound*
x	sy	hs	sh as in *show*
z	dz	ts, tz	ds as in *reads*
zh	j	ch	j as in *judge*

GLOSSARY

acupuncture   A practice used in traditional Chinese medicine in which small needles are inserted into the skin at certain points to relieve pain and cure disease.

autonomous region   One of five large regions of China inhabited largely by people of non-Chinese ethnic backgrounds. Smaller areas within China's 21 provinces are designated as autonomous minor regions, counties, or townships where non-Chinese dwell. Although autonomous means "self-governing," these regions have no political independence. Within them, however, non-Chinese peoples are able to maintain their own languages, customs, and ways of life.

calligraphy   Ornamental handwriting. Chinese calligraphy is done with a brush and is highly prized as an art form.

chuan-chi   Literally, "tale of strange things." A form of short story that included elements of magic or the supernatural.

cloisonné   Enamel applied to metal; an art form that was highly developed in China.

Confucianism   A philosophy or school of ethical thought based on the writings of Kung Fu-tzu (Confucius), who emphasized the importance of order, patience, moderation, service to the state, family loyalty, and obedience.

dynasty   A royal family or house; Chinese history is traditionally described in terms of the ruling dynasties.

Huns   A nomadic people related to the Mongols. They originated in Mongolia and Manchuria, and their aggressive expansion threatened China's northern borders for centuries. The Huns also invaded Russia and Europe.

ideographs   Symbols, commonly called characters, used in writing to represent entire words.

ling   "Mountain range."

loess   A fine, yellowish gray soil that is extremely fertile but also highly subject to erosion by wind and water.

lu-shih   An eight-line poem with an elaborate rhyme scheme.

putonghua   Literally, "common language." The Mandarin dialects of the north, especially the version of Chinese spoken in the Beijing area. This is now China's official national language.

renminbi   "People's money," the official currency of China. The basic unit is the yuan.

shan   "Mountain peak"; also, "mountains," as in Tien Shan.

tai chi chuan   A Chinese martial art that has evolved into a form of noncombative exercise, with emphasis on breathing and balance.

Taoism   A philosophy and religion based on the teachings of Lao-tzu. Taoism emphasizes simplicity, personal harmony with nature, and mysticism. Ritual and magical elements also are present in Taoism.

yueh-fu   A type of poetry that is personal, informal, and can be set to music.

INDEX